Site Surveying
and Levelling 2

Site Surveying and Levelling 2

W. S. Whyte
BA, MSc, ARICS, AIAS, FFB

R. E. Paul
MSST

BUTTERWORTHS
TEC
TECHNICIAN SERIES

First published 1982

© Butterworth & Co (Publishers) Ltd, 1982

British Library Cataloguing in Publication Data

Whyte, W. S.
  Site surveying and levelling 2.
  1. Surveying
  I. Title   II. Paul, R. E.
    526.9′024624   TA549

  ISBN 0-408-00532-7

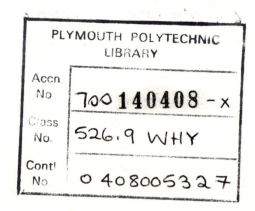
Typeset by Tunbridge Wells Typesetting Services Ltd
Printed and bound by Page Bros Ltd, Norwich, Norfolk

Preface

A student who has worked through the material in this text and has completed appropriate practical exercises, should be able to:

(a) Explain the basic principles of survey and define the commonly-used terms,
(b) Carry out linear surveys using direct linear measurement methods,
(c) Obtain UK national vertical control information and carry out levelling by the traditional methods,
(d) Measure horizontal angles with a modern optical theodolite,
(e) Survey a small building and its site, and
(f) Set out a simple rectangular building and its associated drainage works.

The lecturer familiar with the current Technician Education Council documentation will realise that these general objectives accord with those of the TEC Level II unit 'Site Surveying and Levelling'. However, while generally the text has been planned to meet the requirements of that unit, the book should be suitable for use on any junior technician course in surveying for construction, whether or not that course follows the TEC pattern.

While the range of topics will seem limited as compared with traditional surveying textbooks, each topic is dealt with in considerable detail. The level of detail is such as to allow much of the work to be done by self-study, accompanied of course by supervised practical work. There are certain apparently simple tasks, for example wrapping up a chain, which it is almost impossible to describe effectively in print — we have not attempted this and must rely on lecturers to provide demonstrations of these to clarify the text.

More advanced techniques, such as traverse surveying and the use of co-ordinates, are described in the companion volume *Site Surveying 3,* which similarly also covers the requirements of the TEC level III unit 'Site Surveying'.

Leicester

WSW
REP

Contents

PART A SURVEYING AND LEVELLING

1 BASIC PRINCIPLES
         Introduction   1
1.1   Definitions   1
1.2   Methods of supplying detail, height and control   2
         1.2.1   Methods of supplying detail   2
         1.2.2   Methods of supplying height   4
         1.2.3   Methods of supplying control   7
1.3   Principles of survey   8
1.4   Accuracy, precision and errors   9
1.5   Types of error   9
         1.5.1   Errors caused by carelessness   9
         1.5.2   Errors caused by defective or faulty equipment or methods   10
         1.5.3   Random or accidental errors   10

PART B LINEAR MEASUREMENT

2 DIRECT LINEAR MEASUREMENT METHODS
         Introduction   11
2.1   Equipment and its uses   11
         2.1.1   The chain   11
         2.1.2   The tape   12
         2.1.3   The steel band   13
         2.1.4   Ancillary equipment   13
2.2   Measuring with the chain, tape and band   17
         2.2.1   Standardisation   17
         2.2.2   Laying and using a standard   18
         2.2.3   Tracing or ranging a line   20
         2.2.4   Laying a chain, tape or band   21
2.3   Slope correction in the 'field'   24
         2.3.1   Drop or step chaining   24
         2.3.2   Measuring angle of slope by clinometer   26
2.4   Slope correction — numerical solution   26
2.5   Slope correction — graphical method   27
2.6   Lifting a line   27
         2.6.1   Ranging over a hill   27
         2.6.2   Measuring through a depression   28

2.7   Measuring across or over or through an obstacle   28
         2.7.1   Single 'A' method   29
         2.7.2   Single 'X' method   29
         2.7.3   Inaccessible point method   29
         2.7.4   The 14.02.10 method   30
         2.7.5   Measuring through a rectangular building   31

3 CHAIN SURVEY
         Introduction   33
3.1   Offsets, straights and plus measurements   33
         3.1.1   Offsets   33
         3.1.2   Running offsets   33
         3.1.3   Braced offsets   34
         3.1.4   Straights   35
         3.1.5   Plus measurements   35
3.2   Chain survey booking   36
         3.2.1   The opening page(s) of the field booking   36
         3.2.2   Chain lines, chainage entries and line lengths   37
         3.2.3   Stations, pickets and tie points   38
         3.2.4   Offsets and running offsets   39
         3.2.5   The representation of detail   40
         3.2.6   Braced offsets (ties or tie lines)   40
         3.2.7   Straights   41
         3.2.8   Plus measurements   41
         3.2.9   Referencing detail between chain lines   41
         3.2.10 The continuation of a chain line   41
         3.2.11 Chain line running alongside detail   42
         3.2.12 Detail 'cutting' the chain line   42
3.3   Chain survey equipment   42
3.4   Fieldwork errors   43
3.5   Selection and positioning of chain lines   44
3.6   A practical survey task   45
3.7   Chain survey plotting   45
         3.7.1   Preliminary considerations   45
         3.7.2   Plotting the framework — line plotting   52

3.7.3   Plotting the detail   52
3.7.4   Penning-in — completing the plan   54

PART C   HEIGHT MEASUREMENT

4 BENCH MARKS
    Introduction   58
4.1   Ordnance datum and bench marks   58
    4.1.1   Ordnance datum   58
    4.1.2   Ordnance bench marks (OBMs)   58
    4.1.3   Density of Ordnance bench marks   58
    4.1.4   Types of Ordnance bench mark   58
4.2   The publication of bench mark information   59
4.3   Temporary bench marks (TBMs)   61

5 LEVELLING
    Introduction   62
5.1   The level   62
    5.1.1   The dumpy level   63
    5.1.2   The tilting level   64
    5.1.3   The automatic level   65
    5.1.4   Other levels   66
5.2   The staff and ancillary equipment   66
    5.2.1   The levelling staff   66
    5.2.2   The hand or staff bubble   67
    5.2.3   The staff support   67
    5.2.4   Detachable bracket   67
5.3   Levelling fieldwork   67
    5.3.1   Preliminary tasks   67
    5.3.2   Setting up the level   68
    5.3.3   Reading the staff   71
    5.3.4   Duties of the staffholder   71
    5.3.5   Permanent adjustments of the level   72
5.4   Terms used in levelling   74
5.5   Level booking   75
    5.5.1   The rise-and-fall method   75
    5.5.2   The collimation (height) or height-of-instrument method   77
    5.5.3   Checking levels extending over more than one page   77
    5.5.4   Permissible errors in levelling   78
5.6   Applications of levelling   80
    5.6.1   Flying levels — establishing a TBM   80
    5.6.2   Grid levelling   81
    5.6.3   Levelling for sections and cross-sections   84

5.6.4   Inverse levelling   85
5.7   Sources of error in levelling   85
    5.7.1   Sources of induced error   85
    5.7.2   Sources of instrument error   86

PART D   ANGULAR MEASUREMENT

6 THE MEASUREMENT OF HORIZONTAL ANGLES
    Introduction   87
6.1   The basic components of the theodolite   87
6.2   Types of theodolite   88
    6.2.1   Theodolite reading systems   90
    6.2.2   Theodolite classifications   91
6.3   Field procedure — observing horizontal angles   92
    6.3.1   Setting up the theodolite   94
    6.3.2   Setting a specific reading on the horizontal circle   96
    6.3.3   Observing the direction to a target   97
    6.3.4   Pointing a target with a specific reading set on the circle   98
    6.3.5   Changing instrument station or packing up   98
    6.3.6   Choosing the zero setting to use for a simple reversal measurement   99
    6.3.7   Observing a horizontal angle by simple reversal   99
6.4   Booking horizontal angle readings   100

PART E   BUILDING SURVEYS

7 THE MEASURED SURVEY OF A SMALL BUILDING AND ITS PLOT
    Introduction   101
7.1   Equipment   101
    7.1.1   Equipment for measuring the building   101
    7.1.2   Equipment for measuring the plot dimensions   101
    7.1.3   Plotting equipment   101
7.2   Field procedure on the building site   101
    7.2.1   Preliminaries   101
    7.2.2   The site reconnaissance   102
    7.2.3   Linear measurement   102
    7.2.4   Booking the measurements   104
7.3   Field procedure — measuring the building   104
    7.3.1   Preliminaries   104
    7.3.2   The building(s) reconnaissance   104
    7.3.3   Sketches   104

7.3.4 System used for measuring horizontal and vertical distances 106

7.3.5 Booking methods 107

7.4 Office procedure, plotting 109

7.4.1 Choice of scale 109

7.4.2 Choice of drawing material 109

7.4.3 Choice of sheet size 110

7.4.4 Layout of the survey on the drawing material 110

7.4.5 Plotting the plans 110

7.4.6 Plotting the elevations 111

7.4.7 Plotting the sections 111

7.4.8 Penning-in — completing the drawing 111

PART F SETTING OUT

8 SETTING OUT A SIMPLE RECTANGULAR BUILDING
Introduction 113

8.1 Equipment 113

8.2 Field procedure — locating a proposed building on the ground 115

8.2.1 Stages in setting out 115

8.2.2 Inspecting the documents 115

8.2.3 Inspecting the site 115

8.2.4 Common setting-out techniques 115

8.2.5 Setting out the building plan detail 118

8.3 Field procedure — setting out profile boards 121

8.4 Controlling line and height 122

8.4.1 Controlling line 122

8.4.2 Controlling height 122

9 DRAINAGE WORKS
Introduction and procedure 123

9.1 The sight rail and traveller 124

9.2 Calculations for sight rails 124

9.3 Establishing sight rails 125

Part A — Surveying and levelling

1 Basic principles

Introduction

The object of site surveying and levelling is for one of three purposes:

(a) To produce a map or plan, and/or a section, which can be used in the planning of new 'works' or developments, or

(b) To 'set out' new works, i.e. to place pegs in or other marks on the ground in such a way as to define the location and height of new works or developments which are to be constructed, or

(c) To carry out calculations such as for land areas or earthwork volumes, either direct from the 'field' measurements or from maps, plans and sections.

Over the years, surveying has been classified or subdivided in a variety of ways, and it will be useful at this stage if the reader has an awareness of three of these classifications, as follows:

(a) The classification of surveying as being either (i) 'geodetic' or (ii) 'plane':

(i) *Geodetic surveying* recognises, and makes due allowance for, the curvature of the Earth when attempting to represent an area of land on a flat sheet of paper or other medium.

(ii) *Plane surveying* assumes that the area being surveyed lies on a flat plane, the curvature of the Earth being ignored. Site surveying is most often plane surveying.

(b) The classification of surveys according to their purpose or use, such as:

(i) *Geodetic survey:* A survey of great accuracy which not only takes into account the curvature of the earth but also provides control for surveys of lower accuracy.

(ii) *Topographical survey:* A survey which results in the production of maps and plans showing topography, i.e. the natural and man-made features on the surface of the earth.

(iii) *Cadastral survey:* A survey which results in plans showing and defining property boundaries and, in some cases, is used for land tax purposes.

(iv) *Engineering survey:* A survey made for engineering purposes.

(v) *Mining survey.*

(vi) *Hydrographic survey.*

(vii) *Construction and other site surveys.*

(c) The classification of surveys by the equipment or techniques used, such as chain survey, traverse survey, etc. Some of these will be met later in the text.

1.1 Definitions

presented at scale some suitable form of a map in the form or plan.

Aim: *The student should be able to define surveying, level plane, height, linear measurement and angular measurement*

Surveying may be defined as the art of determining the relative positions of natural and man-made features on the surface of the Earth. In its wider sense the term is generally taken to include levelling operations.

Levelling may be defined as the art of determining the heights of features relative to a datum surface, usually by means of an instrument known as a 'surveyor's level'.

A *datum surface* is any arbitrary level surface to which the height of points may be referred.

A *level surface* (level plane) is one which, at all parts, is at right angles (normal) to the direction of gravity. Since it generally follows the curvature of the Earth it is not a true plane, but over small areas it will be acceptable to regard it as a plane.

Height is the vertical distance of a feature above or below a datum surface.

Linear measurement is the measurement of

distances on a level surface or horizontal plane.

Angular measurement is the measurement of angles in a horizontal and/or vertical plane.

1.2 Methods of supplying detail, height and control

Aim: *The student should be able to explain rectangular offsets, polar co-ordinates, trilateration, triangulation and control*

The following notes provide brief explanations of the methods used in supplying (i.e. surveying) (a) detail, (b) height measurements and (c) control, as used by the site surveyor in the production of a map or plan or in setting out:

(a) *Detail:* The name given in site surveying (and in other large-scale mapping surveys) to the natural and man-made features on and adjacent to the surface of the Earth, but it generally excludes *relief* (heights).

(b) *Height:* Defined in §1.1.

(c) *Control:* A large survey site requires many sets of measurements and calls for methods of tying these measurements together so as to produce an accurate survey. These methods are known as 'control'.

1.2.1 METHODS OF SUPPLYING DETAIL

(a) By (rectangular) offsets
Offsets are short measurements at right angles to a measured straight line. This measured straight line is often known as a *chain line,* hence the method of survey is generally known as *chain survey* and the procedures involved are known as *chaining.* The method is described in depth in Part B, but a basic understanding may be gained from *Figures 1.1* and *1.2* which illustrate the survey of a property boundary line.

The values 3.7, 4.1, 4.9, 5.5 and 4.2 represent the recorded lengths of offsets to points on the hedge where it changes direction and to other changes of the feature.

The values 0.0, 17.1, 32.3, 56.5 and 63.4 are measurements recorded along the chain line to points where it was necessary to raise a right angle to measure the offsets.

The measurements may be plotted and a copy of the boundary line produced as shown in *Figure 1.2.*

In practice, the site of the survey rarely involves only a single chain line and generally requires many chain lines as the offset lengths are usually kept as short as possible (see §3.1). As an example, the survey of a simply-shaped field as shown in *Figure 1.3* would require a minimum of four chain lines to supply the detail. The four lines could be plotted independently but to 'control' the overall shape of the field they must be tied together by, for example, measuring the angles at A, B, C and D or measuring the diagonals AC and BD (*Figure 1.4(a)*) or some other arrangement of triangles as shown in *Figures 1.4(b)* and (*c*).

Note: In all cases these additional measurements or observations not only *control* but also *check* the general shape of the area of survey.

It will be clear, then, that the site to be surveyed must be covered with a network of chain lines which can be plotted to scale and from which all detail can be supplied. If an angle-measuring instrument is not used then the network must essentially be a triangular one.

The choice of method, or combination of methods, will depend upon the size and shape of the land, how it is currently used, whether or not heights are required, the staff, equipment and time available, and usually the cost.

(b) By bearing and distance (polar co-ordinates)
The *bearing* of an observed distant point is the angle between north and the line from the observer's position to the distant point. A bearing, however, represents not merely the angle between the two directions, but also the amount of angular rotation needed to turn from the

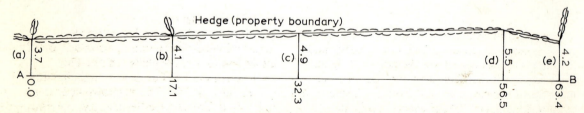

Figure 1.1 (a) to (e) are the offsets from the measured line (chain line) AB

2

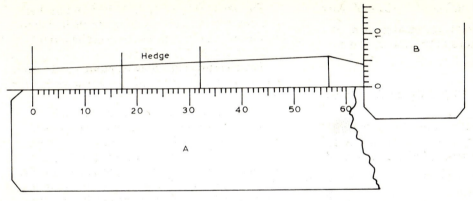

Figure 1.2 Scale positioned A for plotting measured values along the chain line, and B for plotting offset

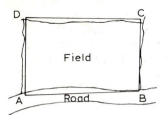

Figure 1.3

 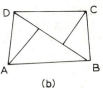 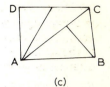

(a)   (b)   (c)

Figure 1.4

northerly reference direction to the direction of the observed point. The rotation is measured in a clockwise direction and the northerly reference direction may be true, grid, or magnetic north.

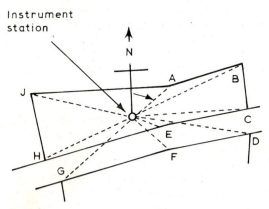

Figure 1.5

Figure 1.5 illustrates how a survey could be carried out to produce a plan by the bearing and distance method. A to J are the points of detail to be surveyed; their bearings must be observed and their distances measured from the instrument station. Plotting of the information may be carried out by the use of a protractor and scale.

The bearing and distance of a point are frequently known as its 'polar co-ordinates'.

Figure 1.6 shows how a site could be set out using this technique, in this particular case using two instrument stations.

As the site becomes larger, whether as regards plan production, setting out, or area calculations, more instrument stations become necessary and hence control must be provided.

The measurement of distances from an instrument station by the traditional (taping) methods can be time-consuming, and the selection of instrument stations can be difficult in 'built-up' areas, making the process sometimes slower than chain survey methods. However, optical distance measurement and, more recently, the introduction of electromagnetic distance measuring equipment suited to site survey use, may make bearing and distance methods more economical.

(c) By intersection
A point of detail may be located by the intersection of a minimum of two lines, one from either end of a known base length. The intersecting lines can be measured from either end of the base line. This is known as fixing a point by *trilateration*. Alternatively, the *bearings* relative to the base line could be measured, the

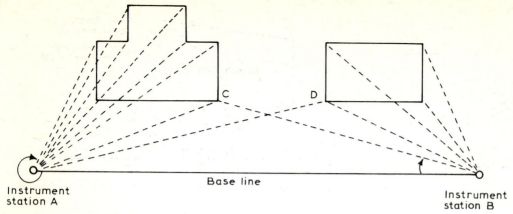

Figure 1.6 C and D are points set out from both instrument stations to act as a check

latter method being known as *triangulation.* The principles may be observed in the checks in *Figure 1.6* and in the Figures following.

Trilateration is illustrated by *Figures 1.7* and *1.8.* The simple method of *Figure 1.7* is often

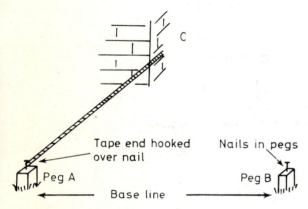

Figure 1.7 At C the surveyor is reading the measured distance to the point to be surveyed

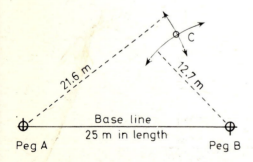

Figure 1.8 Intersection of arcs, C, plotted to scale and hence location of the point on the plan; alternatively, the reverse of this method can be used in setting out

used to measure to points of detail in a small area, single-handed. An example is its use by archaeologists to produce a plan of the permanent antiquities found on a site.

The method is sometimes used in conjunction with chain (offset) survey, further information being given in §3.1.3.

Setting out by this method from two (but preferably three) existing points is acceptable but, when setting out, the tape must never be hooked over the nail.

In producing plans, plotting by trilateration is rather slower than by triangulation but the survey equipment for the latter method is more expensive.

Triangulation is similar to trilateration but the triangle is formed by two bearings (*Figures 1.9* and *1.10*).

Plotting for site surveys is generally by protractor and scale if the angles are measured with a surveyor's compass or theodolite (*see* Part D for surveyors' accurate angle-measuring equipment), but more accurate plotting methods are available. Occasionally, equipment is used in which the angle (bearing) is not read and booked but rather it is plotted direct on the plan in the field. This latter method, plane table surveying, is not dealt with in this book.

1.2.2 METHODS OF SUPPLYING HEIGHT (SPOT HEIGHTS, CONTOURS, GRADIENTS)
(a) By levelling
The commonest method of supplying heights on a site plan (spot heights and contours) or in setting out (including gradients) is by the use of the *surveyor's level* and a *levelling staff.* The staff, which might be described as a giant rule

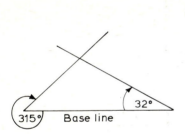

Figure 1.9

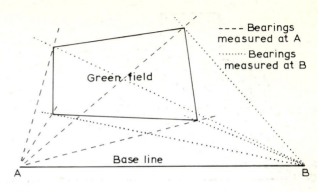

---- Bearings measured at A
.... Bearings measured at B

Green field

Base line

A   B

Figure 1.10

(*Figure 1.11*), is available in a variety of lengths, the commonest probably being 4 m. In use, the staff is held vertically by an assistant and read by the surveyor through a 'level' which is basically a tripod-mounted telescope which may be set up in such a way that all lines of sight through the telescope are horizontal (*Figures 1.12* and *1.13*).

It will be observed from *Figure 1.12* that the telescope height is $25 + 1.3 = 26.3$ m. The surveyor now rotates the telescope in the horizontal plane and an assistant moves the staff onto a peg (*Figure 1.13*).

If the second staff reading is 0.9 m, then the height to the top of the peg is 0.9 m below the horizontal line of sight, that is, $26.3 - 0.9 = 25.4$ m.

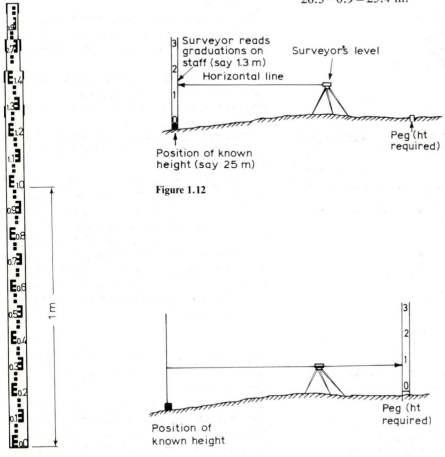

Figure 1.12

Figure 1.11

Figure 1.13

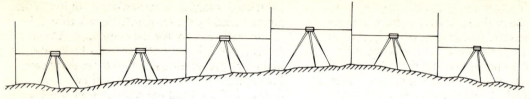

Figure 1.14

There are limitations on the length of sight which may be used, so if it is necessary to transfer heights over a large horizontal distance then the operation has to be repeated (*Figure 1.14*). Part C includes descriptions of the method and the equipment to be used.

barometer which records atmospheric pressure may be used to obtain height measurements. Instead of being calibrated in inches, millimetres of mercury, or millibars, it could be calibrated in metres or feet, and the instrument is then known as an *altimeter*. Anyone who has studied a barometer or a weather

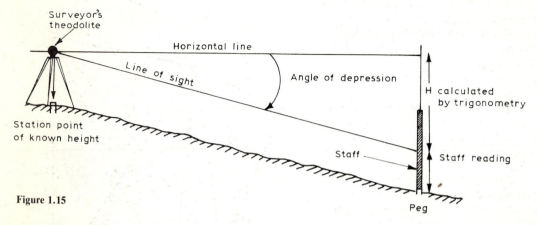

Figure 1.15

(b) By trigonometric heighting

This method, an alternative to levelling, is not generally as accurate, but it may be useful when the ground is undulating or very steep. Instead of a surveyor's level, a theodolite is used and the sight line is directed parallel to the slope of the ground and the staff reading taken (*Figure 1.15*). The theodolite records the vertical angle of the sight line, and, provided that the horizontal or the slope distance has been measured, the difference in height may be calculated by elementary trigonometry.

> Height of peg = height of the station point
> + height of instrument centre
> above the station point
> ± H (calculated height difference)
> − the staff reading.

The mathematics may seem disturbing to some users, but they are simple with a pocket calculator. Further details are given in *Site Surveying 3*.

(c) By barometric heighting

Atmospheric pressure varies with height, thus a

map, will appreciate that atmospheric pressure changes continually *without* any change in height, hence the method is not of sufficient accuracy for site surveying. However, the method is frequently used in many parts of the world for heighting points for small-scale maps. It is not described further in this book.

(d) By hydrostatic (water) levelling

One of the oldest forms of levelling is to use a glass U tube partly filled with water, then sight along the horizontal line (imaginary) joining the tops of the two water surfaces. The glass tube has been replaced by a flexible one with glass or transparent plastic gauges at each end. In site surveying a flexible tube of approximately 30 m in length is used, but tubes of up to 20 km in length have been used in some surveys (*Figure 5.13*). However, it is essential that all air bubbles be removed from the tube and this is difficult over long lengths. Another method, often overlooked, is to use the natural water level of a body of water, provided that the surface is calm and the water flow is negligible.

1.2.3 METHODS OF SUPPLYING CONTROL

Where the site of the survey is extensive it often becomes necessary to carry out additional survey operations for the purpose of controlling the accuracy of the horizontal and vertical measurements. Except for very small sites, most surveys consist of several sets of measurements, such as the measurements of a number of chain lines or sets of observations at instrument stations. These separate sets must be linked together to form an accurate whole, i.e. *control* must be provided in one or more ways.

(a) A base

A *base* or *base line* is simply a straight line of known length extending through or adjacent to the area of the survey. Its length should be approximately the maximum length of the site, but this depends upon site conditions and the methods used to supply detail.

The base controls the scale of the survey, that is, that the completed survey is neither too small nor too large. Hence, the base must be measured with great care using carefully checked measuring equipment. The base also prevents distortion, that is to say it ensures that the completed survey is true to shape.

Examples of the use of a base have been shown in *Figures 1.4* to *1.7* for chain survey. All methods of supplying detail by intersection make use of a base, as in *Figures 1.7* to *1.10* and *1.16.* The base is also used with the other methods of supplying control.

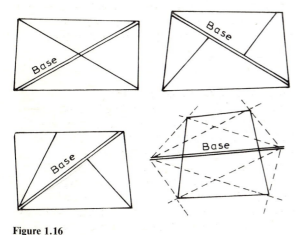

Figure 1.16

(b) Triangulation

Triangulation here means a network of triangles tied to a base of known length, all the angles of the triangles having been measured. This was the method used by the Ordnance Survey to supply the geodetic control for the mapping of Great Britain, and it involved thousands of triangles, an immense computational feat in those early days. Fortunately, in site surveying the network of triangles is small and the computations are few, and occasionally it can be plotted direct without the use of calculations. The method is not covered in this book but an example of the use of triangulation control over a site of approximately 40 hectares for purposes of plan preparation is illustrated in *Figure 1.17*.

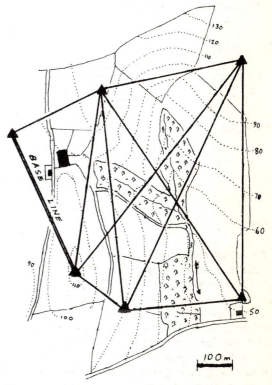

Figure 1.17

The decision as to whether to use this (or any other) method of control will depend upon the site size and shape, current site use, equipment and labour available, methods of detail and height supply, the cost and the time limitations.

(c) Trilateration

Trilateration here means a network of triangles again, but in this case all the side lengths are measured and not the angles. Since accurate traditional linear measurement is very time-consuming over a network of lines, the method was not popular until electromagnetic distance-measuring equipment was introduced. Additionally, control observations must be computed, and this is a very laborious task in trilateration. The method is not covered in this book.

A triangular network of measured lines is used in both chain and building surveys, but these applications are not normally referred to as survey by the method of trilateration.

(d) Traversing

This is the most favoured and most flexible method of providing control for site surveys, particularly in urban areas where the construction of triangles is difficult and sometimes impossible.

A *traverse* consists of a series of connected straight lines whose lengths and bearings can be determined (*Figure 1.18*). The lines are known as *legs* and the end-points of the lines as *stations*.

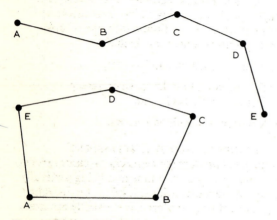

Figure 1.18

The legs may be used either as chain lines for detail supply or for holding a network of chain lines (*Figure 1.19*). Alternatively, detail may be supplied by polar co-ordinates from the traverse stations (detail by bearing and distance). Traverses may commence from and close at either end of a base line (*Figure 1.20*).

See *Site Surveying 3* for traversing as control for site survey.

1.3 Principles of survey

Aims: *The student should be able to explain the principle of 'working from the whole to the part'.*

(a) *Control:* Each survey should be provided with an accurate framework, the lower order work (detail and heighting) being fitted and adjusted to the framework as necessary. The traditional expression is 'working from the whole to the part' and this has been referred to in § 1.2.3.

(b) *Economy of accuracy:* The standard of accuracy aimed for should be appropriate to the needs of the particular task. As a general rule, the higher the standard of accuracy then the higher the cost of the task in time and money — do not work to 0.001 m if 0.1 m is sufficient.

(c) *Consistency:* The relative standards of accuracy should be consistent for each stage of the work (e.g. control, detail, heighting) and should be maintained throughout the whole of the survey area.

(d) *Independent check:* It is desirable that every survey operation (fieldwork, calculations,

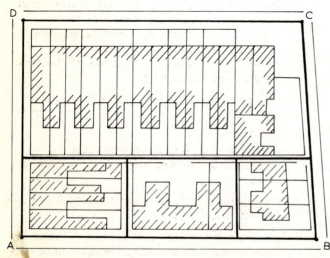

Figure 1.19

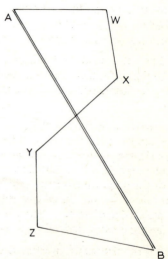

Figure 1.20

plotting, etc.) should be either self-checking or provided with an independent check, e.g. the length of a base line could be measured in one direction in metres and in the other direction in feet.

(e) *Revision:* Surveys should, wherever possible, be planned in such a way that their later revision or extension may be carried out without the necessity of having to carry out a complete new survey.

'Revision' in survey refers to either 'map revision', i.e. new developments on the ground require the map to be brought up-to-date, or, in setting out, further developments planned for an existing site must be set out on the site.

(f) *Safeguarding:* Survey work (survey markers, field and office documents) should be preserved not only for the survey in hand but possibly for use at a later date. For example, a peg should not be placed in the ground when there is a group of children nearby — the peg will not last long!

Do not leave a field notebook overnight on site; by morning it might be soaked by rain or chewed up by the local farmer's livestock! (It is a sensible idea to place your name and address inside the front cover, together with the offer of a small reward for the return of the book.)

Also, take care with modern plotting materials — some plastics materials may shatter or buckle with heat — and ensure that office cleaners do not place your drawing board with attached plotting against or near a radiator.

1.4 Accuracy, precision and errors

Aim: *The student should be able to distinguish between 'accuracy' and 'precision'.*

Standard non-technical dictionaries are seldom clear as to the difference between *accuracy* and *precision*, and some surveyors have the same problem.

Accuracy may be defined as the conformity of a measurement to its true value. The accuracy of a measurement is often quoted as a representative fraction, that is to say as the ratio of the magnitude of the error to the magnitude of the measured quantity. *Error,* here, means the difference between the *true* value and the *measured* value of a quantity.

Precision is the degree of agreement between several measures of a quantity. If a quantity is

measured several times, then the degree of agreement between the measures is the precision of that set of measures. It must be noted that a high degree of precision does not indicate great accuracy. The classic example is the darts player who, while aiming for the 'bull', places all three darts in 'double one' — the player is throwing with some precision (all darts close together) but the result is not very accurate (should have landed in the bull).

1.5 Types of error

Aim: *The student should be able to distinguish between, and explain the characteristics of, different types of error.*

Errors may be of various types, and over the years they have been classified in different ways, authorities on the subject still tending to differ. The site surveyor is recommended to keep the subject as simple and practical as possible by considering errors as falling into one of only three classifications, these being:

(a) Errors caused by carelessness,

(b) Errors caused by defective or faulty equipment or methods, and

(c) Random or accidental errors.

1.5.1 ERRORS CAUSED BY CARELESSNESS

Known as *blunders* (or *mistakes* or *gross errors*), these may occur at any time and being generally large are the most serious, it might be thought. However, if they are large, it is usually apparent that an error has been made, but many hours may be lost in trying to locate the actual error. Frequently it is necessary to make a further visit to the site to re-measure and/or re-observe.

Methods of observing, measuring, booking, computing and plotting should always be designed to show up the occurrence of blunders. Examples of these include:

(a) Measuring a long line and forgetting how many 20- or 30-m tape lengths have been laid down,

(b) Transposing figures when booking measurements, such as writing 12.34 instead of 12.43,

(c) Reading a level staff incorrectly, such as 1.205 instead of 2.205, and

(d) Using the wrong trigonometrical function in calculations, such as using the sine function when it should be the cosine function.

1.5.2 ERRORS CAUSED BY DEFECTIVE OR FAULTY EQUIPMENT OR METHODS

These may have a *constant, cumulative* or *systematic* effect on the results, and they can be eliminated (or at least their effects reduced) by taking appropriate precautions in carrying out the work. Examples of these in linear measurement include:

(a) Errors due to the tape not being of the correct length because it has been either stretched or perhaps broken and then incorrectly repaired,

(b) Errors due to measuring on sloping ground and not reducing the measurement to the horizontal equivalent distance,

(c) Errors due to applying the wrong tension to the tape,

(d) Errors due to allowing the tape to sag when measuring across an open drain, and

(e) Errors due to allowing the tape to become 'bowed' out to the side by a strong wind blowing across the line being measured.

Error (a) is systematic and it will always occur when that particular tape is used. Its effects will be cumulative since the longer the line measured then the greater the error will be. In this case, if the tape is *too long*, then a *positive* correction will have to be applied, but if it is *too short* then a *negative* correction will be required.

1.5.3 RANDOM OR ACCIDENTAL ERRORS

Errors which remain after the errors described above have been eliminated or reduced are known as *random* or *accidental* errors. These are small, and as they may tend to compensate one another if some are positive and some are negative, they are not cumulative in the general sense. They may be due to small imperfections in equipment, to changing environmental conditions, or to minor discrepancies such as the surveyor possessing a personal bias such as a tendency always either to underestimate or overestimate the value of some reading.

In practice, these small errors and some cumulative errors will remain on completion of the measurements, and in a control network, in plotting and in setting out they must be distributed so as to minimise their effect on the final result.

Part B — Linear measurement

2 Direct linear measurement methods

Introduction

Linear measurement has been defined in Part A as the measurement of distance on a level surface or on a horizontal plane. This book deals only with linear measurement as carried out by the use of a chain, tape, or steel band, as described in §§ 2.1.1, 2.1.2 and 2.1.3 respectively. This form of linear measurement is often known as *traditional linear measurement*, or *direct linear measurement* but lines may also be measured optically, or with electromagnetic distance-measuring equipment, or by pacing, using a road measuring wheel, etc.

Traditional linear measurement is used in chain surveys, in building surveys, and in setting out, and all of these are covered in this text.

There are two essential requirements in linear measurement, the first being that a straight line be maintained, the second that an accurate measurement be obtained along the line. The former contributes to the latter, and the standards to be observed in these will depend upon the objective and the type of survey to be carried out. Techniques will therefore vary according to the accuracy demanded, the type of topography and the equipment available.

See also linear measurement: *in building surveys* — Part E; *in setting out* — Part F and *Site Surveying 3; in traversing* — *Site Surveying 3*.

2.1 Equipment and its uses

Aim: *The student should be able to identify chains, tapes and bands, and to state their uses.*

2.1.1 THE CHAIN

Several forms of chain are available, according to the unit of measurement in use. The form illustrated in *Figure 2.31* is the 20-m chain specified in BS 4484 of 1969, which is the type recommended by the authors for carrying out 'chaining'.

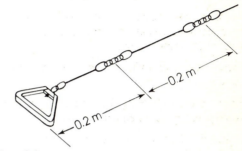

Figure 2.1

The chain comprises 100 steel wire 'links', each of which is joined to its neighbours by two or three oval rings. Swivelling brass handles are fitted to each end of the chain and its total length of 20 m is measured from the outside of one handle to the outside of the other. It may be noted that the nominal length of the chain is stamped on the handles. *Figure 2.1* shows a part of such a chain.

To enable the chain to be read with ease, yellow plastic 'tallies' (*Figure 2.2*) are fitted at

Figure 2.2 Yellow tally (actual size)

every metre except at 5, 10 and 15 m where red numbered tallies are used (*Figure 2.3*). **Note:** There are 5-m tallies at both the 5 and 15 m positions, thus when laid on the ground the chain may be used to measure in either direction.

An individual reading of a 'chainage' figure may be read by estimation to 0.05 m. A line measured with a chain is, at its best, possibly accurate to 1:1000, that is to say correct to one metre in a thousand metres measured or to 0.1 m in every 100 m measured. In practice, however, chain measurements are often correct to only 1:500 or less, thus it is obvious that the chain is not a very accurate measuring instrument. Nevertheless, it will be seen later that it is adequate for the majority of chain (offset) surveys and even occasionally for setting out tasks where high accuracy is not required.

Figure 2.3 Red tallies (actual size)

The advantages of the chain are that it is robust, vehicles may drive over it and the chain is rarely damaged. Due to its own weight it will, when correctly used, lie at ground level in fields of stubble, long grass and long weeds. On the other hand, it may damage plants if dragged through gardens and it cannot lie in a flat plane if the ground is very broken as on disused industrial sites. It should also be noted that some links may be bent by heavy vehicles passing over them, and mud, snow and ice may lodge between the small oval rings. The effect of any of these is to shorten the length of the chain, hence the site surveyor must beware of these possibilities and check as necessary.

2.1.2 THE TAPE

Tapes are manufactured either from steel or from strands of glass fibre coated with PVC, the latter often being referred to as 'synthetic' tapes. Steel tapes tend to rust and to break easily, but these disadvantages have been partially overcome by the use of stainless steel or by using enamel or plastic coatings on ordinary steel tapes. A variety of lengths up to 100 m are available, but lengths of 10, 20 or 30 m are generally preferred for site surveying. The recommended length for a particular job will be referred to later.

Tapes are made in a range of widths but 10 and 13 mm are most favoured. They are usually fixed into and carried in a case of plastic, leather or steel with a recess to accommodate the folding winding handle. The zero mark is at the outside of the end ring (*Figure 2.4*). Graduation patterns differ, and some site surveyors have particular preferences — what is essential, however, is that the graduations should be in the desired unit (e.g., 1 mm, 10 mm, etc.) and easily readable without ambiguity. Examples of tape graduations are shown in *Figure 2.5*.

Measurements can be made much more accurately with a steel tape than they can with a chain. Using accepted procedures an accuracy of 1:2000 (50 mm in 100 m) is easily attained and still better results can be achieved, but it is generally advisable to use the steel band if higher accuracies are demanded. Greater accuracy than 1:2000 is often desirable for control surveys and occasionally in setting out when the allowable error is small and there are long distances to be taped over. The steel tape is adequate, but not always desirable, for measurements to detail for plan production, in building surveys, and for the majority of setting-out tasks.

The advantages which the steel tape has over the chain are, of course, its greater accuracy, its lighter weight, and its lack of bulk. Its disadvantages are that it rusts easily, hence it must be cleaned and dried after use, and non-rustless tapes must be wiped with an oily rag after use. Steel tapes kink easily and this frequently leads to their breaking. Even a bicycle passing over a steel tape laid flat and with no tension applied may break the tape. Any loop in a steel tape should be unwound by hand, since any attempt to pull the loop out will usually cause a kink and lead to a break. A tape's light weight is,

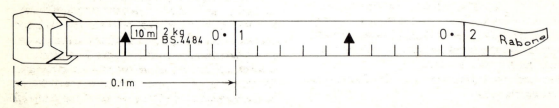

Figure 2.4

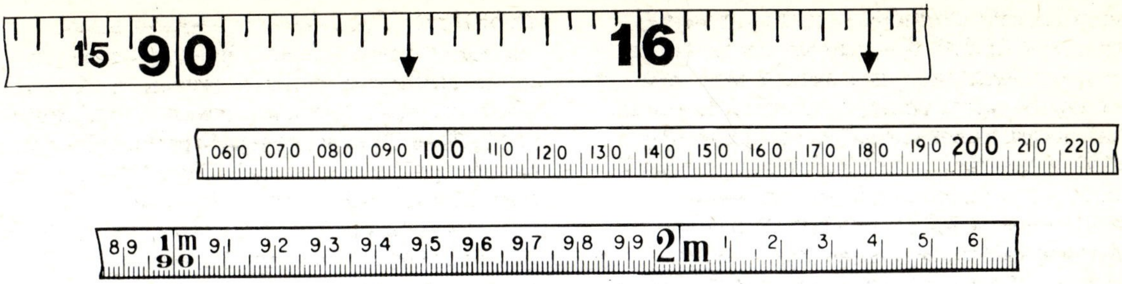

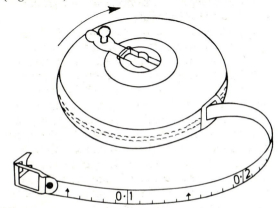

Figure 2.5

of course, an advantage from the point of view of carrying it or measuring over broken ground, but it is a disadvantage under windy conditions or when measuring in long grass, tall weeds, etc.

It is generally considered that a synthetic tape (*Figure 2.6*) is not as accurate in use as the chain,

Figure 2.6

and it should therefore be used only for short measurements such as offsets in chain survey. It cannot be recommended for use in setting out except in very rough work. An advantage is that it is cheaper than steel tape or chain and in some ways is easier to use, but it has a much shorter life than has a steel tape which has been treated with care. Synthetic tapes may safely be used in the vicinity of electric fences and railway lines and transmission lines, provided that the tape is kept dry.

2.1.3 THE STEEL BAND

As a piece of linear measuring equipment the steel band is basically no more accurate than the steel tape, but the greater accuracy which can be achieved with the band is due to the particular measuring procedures used.

The steel band differs from the steel tape in the following respects:

(a) The *minimum* length is usually 30 m.

(b) It is narrower, being approximately 7 mm in width.

(c) The zero mark is etched on the band at about 150 mm from the end ring.

(d) The end ring is often replaced by a detachable handle similar to a chain handle, or sometimes by leather thongs.

(e) The band is detachable from its case since in use this is often preferable if greater accuracy of measurement is to be achieved without damaging the band.

(f) Generally the carrying 'case' is actually an open winding frame, for easier removal of the band.

(g) The winding handle serves as a locking device for retaining the band on its frame.

Figure 2.7 shows a typical modern steel band in its winding frame.

2.1.4 ANCILLARY EQUIPMENT

Earlier sections referred to the need to obtain a straight line and to arrive at an accurate measurement of distance. To achieve these aims a range of ancillary equipment is needed in addition to the tape or band, and these items are described here. (The theodolite is of importance in higher accuracy work — this is dealt with in Part D and in *Site Surveying 3*.)

(a) Arrows
The *chaining arrow* is a steel wire pin, roughly 0.35 m in length (*Figure 2.8*) which is used to mark the end of a chain or tape length laid down. When great accuracy is necessary, or if an arrow cannot be used (e.g., on a concrete surface), then a fine mark is drawn, cut or scratched on a smooth surface. Such surface may be achieved, for example, on a rough surface, by pressing on-to it a thin layer of plasticene on which a straight cut can be made with a penknife to indicate the

Figure 2.7

chain or tape end. In rural areas, mud forms a suitable substitute.

Arrows are also used to record the number of chain or tape lengths laid down when measuring a line. One of the commonest blunders when measuring long lines is to forget exactly how many chain or tape lengths have been laid. This may be overcome by always using a constant number of arrows, often ten, and carefully checking procedures.

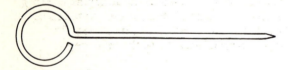

Figure 2.8 The chain arrow

The typical 'chaining' of a line requires a team of two, a surveyor and an assistant, the latter pulling the tape or chain forward while the surveyor 'drives' him from the rear. The surveyor lines in the assistant at the far end of the tape or chain, ensuring that he is holding the foward end correctly on the line which is being measured, then the assistant places an arrow in the ground at the 20-m mark (assuming the chain is 20 m), or alternatively he makes a chalk mark close to which he lays the arrow on the surface. When the chain is pulled forward again to lay down the *next* length, the surveyor walks forward and picks up the arrow. Since he starts at the beginning of the line with *no* arrows, and the assistant starts with *ten* arrows, the number of

arrows in the surveyor's hand will always indicate how many chain lengths have been laid.

At the end of the chain line, or at any point along that line, the value on the particular chain length is read and to it is added the distance indicated by the number of chain lengths shown by the number of arrows in the surveyor's hand (often hung on the little finger). From time to time the surveyor should check the number of arrows held by both to ensure that none have been lost, since a lost arrow may cause incorrect recording of distance measured. When the line has been measured, all arrows should be returned to the assistant before starting to measure the next line. *Figure 2.9* shows a surveyor reading an offset of 2.5 m to the corner of a building, at a chainage value (distance along the chain line) of 65.1 m.

(b) Ranging rods or poles

These are one- or two-piece poles of wood or metal, pointed at one end and made in various lengths but typically 2 m (*Figure 2.10*). They are painted in bands of 0.2 or 0.5 m width, alternately red and/or black and white. They are useful for marking points on lines and the ends of lines to be measured. The preferred form is the 2-m rod with 0.2-m banding, this type being very useful, in addition to its main function, for measuring short offsets between the chain line and detail.

(c) The optical square

Several types of optical square are made, the best form having two pentagonal prisms mounted one

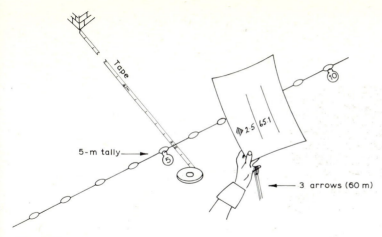

Figure 2.9

above the other in a metal or plastic housing (*Figures 2.11* and *2.12*). The instrument is used for checking and establishing alignments, and it will be seen later that it is also particularly useful for 'raising' offsets in chain survey.

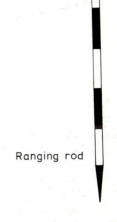

Ranging rod

Figure 2.10

(d) The Abney level or clinometer
The clinometer is a hand-held instrument, used to observe the angle of slope of the ground along which a straight line is to be measured. The *Abney level (Figure 2.13)*, the most popular type

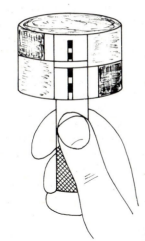

Figure 2.11

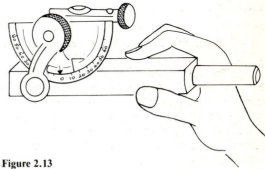

Figure 2.13

of clinometer, comprises a rectangular sighting tube, a graduated semi-circular vertical arc with vernier scale, a bubble tube and a mirror. The mirror mounted in the tube enables the bubble to be observed in coincidence with the object being viewed (*Figure 2.14*).

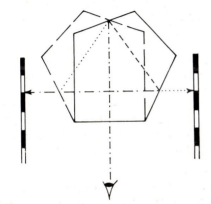

Figure 2.12

15

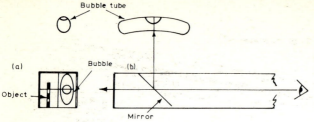

Figure 2.14 Sections: (a) through, and (b) along, a sighting tube

The semi-circular arc is graduated in degrees, being read to 5 or 10 minutes of arc by estimation, using the vernier scale and a magnifying glass (*Figure 2.15*). The instrument is not suitable for use where slopes are excessively steep or where high accuracy is demanded, but it is appropriate for surveys of moderate accuracy on gentle slopes.

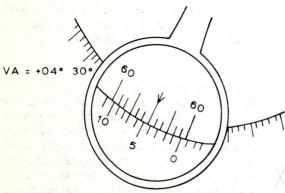

VA = +04° 30°

Figure 2.15

Note: Most Abney levels also have a gradient scale on the graduated arc.

(Remember that in site surveys distances are required, or given, as if measured on a level surface or horizontal plane, and distances measured on slope must be corrected to allow for the slope.)

(e) The thermometer

Materials expand and contract with changes in temperature so for more accurate work the temperature of the steel tape or band is required. The 'field' temperature is normally recorded by the use of a simple thermometer which should be capable of covering the full range of working temperatures, and should be provided with a protective case or sleeve.

The coefficient of thermal expansion for the majority of steel tapes and bands may be taken as 0.000 011 per degree Celsius, i.e., for every rise/fall of temperature of one degree Celsius the steel tape will expand/contract 0.000 011 metres for each metre of its length. To look at it in a more practical way, a 100-m band will vary in length by approximately 22 mm between being used on a very cold day and on a warm day, or again between a warm day and a very hot day. This figure of 22 mm is arrived at as follows:

Correction to tape length = difference in temperature × coefficient of thermal expansion × approximate length of the band,

or, $c = \Delta t \times 0.000\,011 \times L$,

Therefore, if the temperature on a very cold day is 0°C, and on a warm day 20°C, and the measured length is 100 m,

then $c = 20 \times 0.000\,011 \times 100$
$= 0.022\ m$

(c = correction to measured length, Δt = difference in temperature, 0.000 011 = coefficient of thermal expansion, L = approximate length of tape or band.)

With increasing familiarity with the subject, it will become evident that the recording of temperature, and the need to make allowance for it, are required only in control surveys (§ 1.2.3) and occasionally in setting out.

(f) The spring balance

Again in control surveys, and occasionally in setting out, the tape or band has to be laid out with the correct tension applied and to do this successfully a spring balance is used (*Figure 2.16*).

Practical tests have shown that failure to use a spring balance can result in readings being incorrect by as much as 15 mm per 30 m (i.e., 1:2000). Hence, if accuracies of greater than 1 in 2000 are required it is preferable to use a spring balance.

(g) The tape grip

To enable the spring balance to be attached to any point along a steel band, rather than merely

Figure 2.16

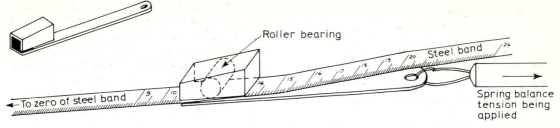

Figure 2.17 Littlejohn tape grip

hooked into the band end ring, the 'Littlejohn' roller tape grip may be used (*Figure 2.17*).

(h) The hand bubble
This is a small hand-held bubble, useful for setting (plumbing) a ranging rod in a vertical plane (*Figure 2.18*).

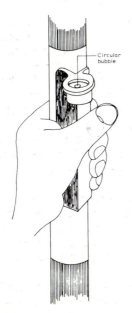

Figure 2.18 The circular bubble is attached to a metal or plastics angle bracket which is held by the hand against the ranging rod

2.2 Measuring with the chain, tape and band

Aim: *The student should be able to demonstrate the uses of chains, tapes and bands, and compare their accuracies.*

Different procedures are followed in chain surveys, building surveys, and in setting out, but the ability to keep the measured line straight and obtain an accurate measured distance along this line is essential in all cases. The following sections consider these three pieces of measuring equipment and also whether the distance to be

measured is less or more than one chain, tape or band length.

Prior to commencing actual measurement, two tasks may have to be carried out. The first is known as *standardisation*, and this involves comparing the chain, tape or band to be used with some standard of known length. The second task is to *trace* or *range the line* (§ 2.2.3) which requires a series of marks to be placed on the ground at intervals along the line which is to be measured. Ranging the line assists the site surveyor in maintaining a straight line when measuring.

2.2.1 STANDARDISATION

All measuring equipment must be checked from time to time against some standard. This might be a steel tape reserved for this particular purpose or, preferably, an accurate standard laid down in, or adjacent to, the site surveyor's place of work or office. The simplest form of standard, which has an accuracy sufficient for the majority of site surveys, requires the following equipment:

(a) A standardised steel band which might be one reserved from new and used solely for standardisation. Either this has etched on it the temperature and the tension to be applied at which the graduations on the band will be correct, or there may be issued *with* the band a certificate on which the essential information has been recorded. *Figure 2.19* shows a copy of a certificate for a 100-m band, issued by the Yamayo Measuring Tools Co. Ltd., Tokyo, Japan.

Note: Bands issued without a certificate, although possibly new and said to be correct (e.g. at 20°C and under 7 kg tension) may in fact be accurate to only ± 2.5 mm per 30-m length.

(b) A spring balance, which should have previously been compared with another as an accuracy check. (Useful to keep a tape grip with the balance.)

(c) Two or three thermometers.

(d) Equipment for making permanent or temporary marks. Marks may be simple crosses drawn with ball-point pen on masking tape, or they may be brass rivets either marked with a fine saw cut or centre-punched. Other alternatives may be considered, such as steel scales fixed to a concrete floor with adhesive.

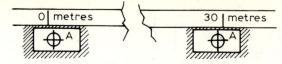

Figure 2.20 A is a cross marked on masking tape with ball-point pen

Inspection Record List			
Description	Tough Stainless Tape		
Length	100m/330ft	Temperature	20°C
Tape No.	A-2624	Tension	~~2kgf~~ ~~5kgf~~ 6.8 kgf (15 lb) ~~10kgf~~
Length	Real length (m)		
0~5 m	4.77990		
0~10	9.99970	Accuracy of inspection :	
0~15	14.99960	±0.1 mm	
0~20	19.99960	Date of Inspection	
0~25	24.99960		
0~30	29.99960		
0~35	34.99935	24TH JAN. 1972	
0~40	39.99925		
0~45	44.99930	Should there be any question	
0~50	49.99930	about the mentioned	
0~55	54.99945	allowance, please contact us	
0~60	59.99955	with Tape No.	
0~65	64.99970		
0~70	69.99960		
0~75	74.99970	YAMAUO MEASURING TOOLS	
0~80	79.99970	CO. LTD	
0~85	84.99970	4-2 Nihombashi, Honcho.	
0~90	89.99975	Chuo-ku, Tokyo, JAPAN	
0~95	94.99985	N. Kamoshita	
0~100	99.99987	Inspection Dept.	

Figure 2.19

2.2.2 LAYING AND USING A STANDARD

This section describes how to lay a standard suitable for the comparison of 'field' bands, when an accuracy of 1:10 000 may be necessary (e.g. 3 mm in 30 m). The technique may be modified when this accuracy is not required, and it should be noted that this accuracy is only very occasionally necessary for any of the site surveying tasks referred to in this book. The standard to be laid, often known as a *standard bay*, can be the approximate length (or some fraction of it) of the steel tape or band. As an example, a 50-m band may be standardised using a 25-m standard bay.

The standard tape or band should be laid out on flat ground where there is little or no vehicular traffic. There must be permanent features at the terminal points of the bay on which the bay length may be marked. With the band lying on the ground, approximately straight and without tension, two marks should be placed close to the zero and, say, the 30-m mark (*Figure 2.20*).

The band must now be straightened, without twists, and aligned over the marks by shaking

(snaking) it in a vertical plane. It is usually sufficient to snake the band gently from one end only while the other end is held in the approximate position with a hand or foot.

The spring balance should be attached at the 30-m end of the band, and one or two thermometers laid on the ground adjacent to the band but well clear of the site surveyor and assistants (A team of three is the minimum for this task but other assistants may be needed if vehicular or pedestrian traffic is liable to damage the band.) In carrying out the standardisation, one assistant lays the zero mark of the band on the zero mark of the bay, the second assistant applies the required tension, and the surveyor reads the graduations on the band at the 30-m end of the standard bay.

On the command READY being given by the site surveyor, the zero mark of the band must be held exactly on the zero mark of the bay and the correct tension applied with the spring balance at the 30-m end. Initially, there will be some longitudinal movement of the band until the assistants have appreciated the effort required to keep the correct tension. When the band has steadied, the assistant at the zero end of the band shouts ON as soon as that end is correctly positioned. The assistant at the 30-m end then whispers *on* when the correct tension is being applied. The surveyor should be kneeling on the ground ready to read the band, and on hearing ON, *on* he should read and record the value of the band graduations against the end (30-m) mark of the standard bay, and finally shout RELAX.

This last order allows the tension to be eased off until the surveyor gives the command AGAIN, when the whole process is repeated to obtain another reading. It is normal practice to read the length of the band being standardised against the bay at least three times, and the three readings should agree within a range of 3 mm. Finally, the thermometers are read and their temperatures recorded. The process is illustrated in *Figure 2.21*.

A *field tape* or *band* may be checked against the standard bay by using exactly the same process.

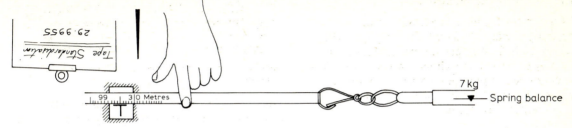

Figure 2.21 With his finger on the tape applying minimum pressure to keep the band flat on the ground, the surveyor reads and records the value on the tape at bay mark

Figure 2.22 shows sets of readings obtained from the standard bay using both a standard band and a field tape. From these readings it will be seen that the mean observed length of the standard bay, using the standard band, is 29.9758 m, at 18°C and 6.8 kg (15 lb) applied tension. If the temperature had been 20°C the standard band would have expanded and this would have caused the standard bay to be recorded as a shorter length. The difference in length may be calculated from the formula:

$$c = \Delta t \times 0.000\,011 \times L$$
$$= (t - 20) \times 0.000\,011 \times L,$$

in this case

$$c = (18 - 20) \times 0.000\,011 \times 30$$
$$= -2 \times 0.000\,33$$
$$= -0.000\,66 \text{ m},$$

or, −0.0007 m to four decimal places.

Thus, at 20°C the length of the standard bay, measuring with this particular standard band, would be recorded as 29.9758 − 0.0007 = 29.9751 m.

However, as *Figure 2.19* shows, the Yamayo certificate for this standard band records that at a distance of 30 m as shown on the tape the actual tape length is 0.0006 m less (at 20°C and standard 6.8 kg (15 lb) tension). Accordingly, the calculated length of the standard bay must be reduced by 0.0006 m, and the length of the standard bay will be taken as

$$29.9751 - 0.0006 = 29.9745 \text{ m}$$

When a bay is established outdoors, the calculated bay length is generally taken to be the same at all temperatures, although this may not, in fact, be the case. If an indoor bay is used, some correction may be required to allow for thermal movement of the material upon which the standard bay is placed. As an example, the coefficient of thermal expansion of timber is 0.000 005 per degree C.

Using a standard is illustrated by the readings for Field Tape No. 5 in *Figure 2.22,* where the standard bay was found to measure 29.9705 m with Tape 5 at a temperature of 16°C and at its correct tension of 7 kg. At 20°C the field tape would have expanded, hence the length would have been recorded as being shorter by

$$(16 - 20) \times 0.000\,011 \times 30 = -4 \times 0.000\,33$$
$$= -0.001\,32$$

or, −0.0013 m to four decimal places. Thus, at

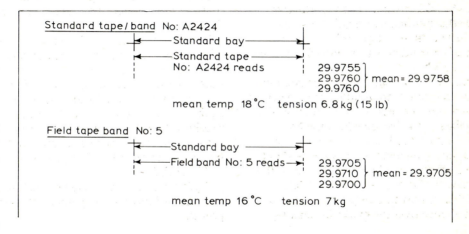

Standard tape / band No: A2424

```
      |←————Standard bay————→|
    |←————Standard tape————→|
    │     No: A2424 reads    │   29.9755⎫
    │                        │   29.9760⎬ mean = 29.9758
    │                        │   29.9760⎭
```

mean temp 18°C   tension 6.8 kg (15 lb)

Field tape band No: 5

```
    |←————Standard bay————→|
    |←——Field band No: 5 reads——→|   29.9705⎫
    │                             │   29.9710⎬ mean = 29.9705
    │                             │   29.9700⎭
```

mean temp 16°C   tension 7 kg

Figure 2.22

20°C the length of the standard bay would have been recorded as

$$29.9705 - 0.0013 = 29.9692 \text{ m}$$

and for every 30 m of taping there would be, at 20°C, a difference of

$$29.9745 - 29.9692 = 0.0053 \text{ m}$$

between the length shown by Tape 5 and the correct length.

This 0.0053 m correction may be used in one of three ways, according to choice:

(a) State the correction for standard as +0.0053 m per tape length (or some other length, e.g. as +0.0177 m per 100 m measured); or

(b) Use a correction factor, if a calculator is available. The correction factor f is the quantity by which the length of the field tape or band (29.9692 m here) must be multiplied in order to equal the correct length of the standard bay (29.9745 m here). In this case,

$$29.9692 \times f = 29.9745$$
$$\therefore f = 29.9745/29.9692$$
$$= 1.000\,18 \text{ m}$$

and this is the method recommended by the authors.
Alternatively

(c) Determine at what temperature the field tape must be so that when correctly tensioned it will read exactly 30 m. Considering the same example, *Figure 2.23* shows that the bay length is

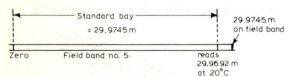

Figure 2.23

fixed and the field tape must be 'shrunk' so that the 29.9745-m mark on the tape coincides with the end mark of the bay, which is known to represent a true length of 29.9745 m. The temperature change required for this to occur may be obtained from the previous formula:

$$c = \Delta t \times 0.000\,011 \times L$$
$$\text{or, } \Delta t = c/(0.000\,011 \times L)$$
$$= 0.0053/(0.000\,011 \times 30)$$
$$= 0.0053/0.000\,33$$
$$= 16°C$$

Hence, the field tape would read correctly at

$$20 - 16 = 4°C$$

How to use this information is explained in § 2.2.4(c).

2.2.3 TRACING OR RANGING A LINE

It is generally essential that there be no deviation from the straight when measuring a line. Consequently, lines of more than one chain, tape or band length may need to be aligned, i.e., traced or ranged, before measurement. This is done by placing marks such as ranging rods, or sticks, or sticks with pieces of paper impaled on them, at intervals along the line to be measured. Even the most experienced site surveyor might be liable to 'bend' a line by as much as 0.3 of a metre if the alignment has not been fixed prior to measurement.

It is recommended that line ranging be carried out in the following circumstances:

(a) For all lines exceeding 200 m in length, unless the alignment is held by points of detail such as electricity poles, gateposts, lamp posts, lines of straight walls or kerbs, etc., and

(b) For all lines which must be measured over very broken ground such as industrial waste land, fenced or walled yards and gardens.

In site surveys where lines of over 400 m length are to be ranged, it is preferable to use a theodolite for the task, but this section will deal only with direct ranging by eye. (The theodolite is described in Part D, *Angular Measurement*, and its use for lining-in between points is explained in Part F, § 8.2.)

When a line is to be ranged, the site surveyor must first decide where the line is to start and mark that point. He must then either mark the end point or, if that is uncertain, place a mark on the line which will fix its direction. (Often, in chain survey, it is convenient to select a 'forward object' such as a building corner, or a vertical pole or post which is on the required alignment but is well beyond the anticipated end of the line.)

The surveyor stands over the selected starting point mark and directs the assistant to proceed along the proposed line for about 100 paces, then turn and face the surveyor, holding a ranging rod vertically. The surveyor signals the assistant to move the rod until it is on the correct alignment, and on receiving the signal 'on line, mark' the assistant marks the position. On soft ground, the mark may be a ranging rod, a peg, or stick and

Figure 2.24 'Assistant, move to your right'

Figure 2.25 'On line, mark'

paper, otherwise a chalk or crayon mark can be used. This procedure is repeated at intervals of not more than 100 m until the end of the line is reached. In chain survey it is desirable to ensure that similar marks are placed on the line in the vicinity of 'tie points'. A tie point is a point on a chain line at which another chain line commences or closes (*Figure 2.26* and § 3.2.1).

(a) Laying a chain
The line having been ranged (if necessary) the chain is 'thrown out' or dragged in the direction it is desired to measure. The technique is for the surveyor to unwind a few links and, holding the handle(s) in one hand, throw the chain bundle forward along the line. He then stands on one handle while the assistant pulls the other handle forward along the line. The chain must be straightened by gently pulling and snaking it, and when it is straight the surveyor and assistant should inspect it for any tangled links which must be untangled.

The outside of a handle is placed at the start of the line and retained in position by the surveyor (*Figure 2.27*) while the assistant at the other end

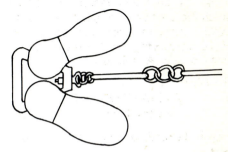

Figure 2.27

holds the other handle and faces the surveyor. The assistant should hold the ten arrows in the other hand or, preferably, in a small canvas bag or similar holder. The surveyor then directs the assistant onto the correct alignment and, when approximately on line, the assistant crouches

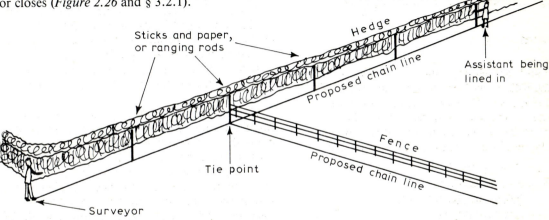

Figure 2.26

21

while holding the chain with sufficient tension to keep it straight (*Figure 2.28*). The surveyor looks along the chain and 'through' the assistant's hand holding the chain to check correctness of the alignment, signalling to the assistant as necessary. Straightness and movement onto the correct alignment are achieved by the assistant snaking the chain and moving as needed slightly to one side or the other, then on the signal 'On line, mark' the assistant lowers his handle to the ground. The surveyor repeats the signal if the alignment is correct, otherwise the chain is re-aligned.

Figure 2.28

On receiving the second 'On line, mark' signal the assistant releases the handle so that the chain is no longer under tension. An arrow is then inserted in the ground against the chain handle, or on hard ground a chalk T mark is made and the arrow laid on the ground close to it (*Figure 2.29*). **Note,** however, that if the chain lies across hollows then some tension will be necessary. The surveyor should remain standing on the zero end of the chain until the far end has been adequately marked.

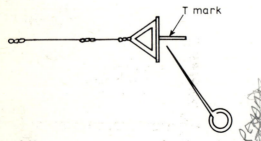

T mark

Figure 2.29

With the first length of any chain line laid, the assistant selects a 'back' object beyond the surveyor before either moves. This allows him to position himself on the approximate alignment for each succeeding chain length (*Figure 2.30*).

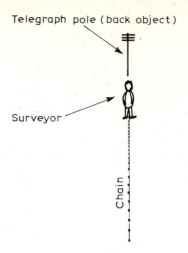

Telegraph pole (back object)

Surveyor

Chain

As seen by assistant

Figure 2.30

When ready to move forward again the surveyor calls NEXT CHAIN and the assistant drags the chain forward using the back object as a guide. When the assistant still has about a couple of metres to drag the chain the surveyor shouts CHECK!, although a good assistant should be aware of the number of paces in a chain length and know when to stop. The process is repeated for laying the second and subsequent chain lengths.

The final chain, or part chain length, is laid in a similar manner but the assistant should not need to be placed on line if the end point of the line has been previously marked.

The surveyor collects the arrow at the end of each chain length laid, but only after the subsequent arrow has been laid or inserted. These collected arrows the surveyor hangs on his finger (*Figure 2.9*). If the length to be measured in one line is greater than ten chain lengths, then a nail or peg should be inserted in the ground at the end of the tenth chain and all the arrows returned to the assistant to be used again.

The chain is read to the nearest 0.05 m at best, and if there is any doubt in the mind of the surveyor as to whether a length is, say, 72.35 or 72.40 m, then the lower value reading should be accepted. This is because in practice the length required is always less than the measured length, since very few lines are truly horizontal or on a level plane.

If, at any time, the site surveyor considers that the correct alignment is not being maintained, then it should be checked with the optical square as described in § 2.6.1. Short measurements of

Figure 2.31

less than one chain length are, of course, similar in practice to laying the final chain length or part chain length of a longer line.

The chain is usually dragged from line to line, being 'wrapped up' only at the end of the day's work. The 20-m chain may be wrapped up from one end in 'concertina' fashion, the bundle finally being tightly secured with a strap or sling passed through the handles, forming a 'wheatsheaf' (*Figure 2.31*).

(b) Laying the tape

To lay out the tape, the surveyor holds the zero end while the assistant, holding the tape box behind him with the winding handle folded inwards, walks in the direction of the line to be measured, allowing the tape to unwind as he goes. Again, knowledge of the number of paces needed is desirable to avoid the tape box being jerked out of the assistant's hand. The tape must be laid down straight and free from twists along its length, with the zero (outside of handle or ring) on the start point. The technique is similar to that for laying the standard tape/band or the chain, but the site surveyor may have to assist in aligning the tape by 'snaking' at the zero end, particularly if the ground is stony or covered in vegetation. Further, the surveyor must adopt a different stance due to the smallness of the tape handle (*Figure 2.32*).

On some occasions it is preferable to hold the tape with the hand (*Figure 2.33*), but as in all cases in lining-in the assistant, the surveyor's eyes should be vertically above the starting point of the line. On some occasions, for example on soft

ground, it may be preferable to align the tape first with the tape zero on the correct alignment but not necessarily at the starting point of the line. When correctly aligned, the tape may then be moved along the alignment until the tape zero is over the starting point. Finally, when the tape

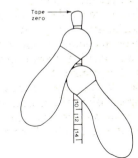

Figure 2.32

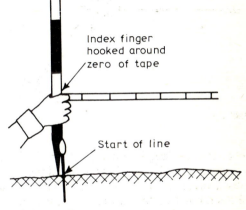

Index finger hooked around zero of tape

Start of line

Figure 2.33

is aligned and correctly zeroed, the surveyor signals 'on line, mark' and the assistant, applying and maintaining sufficient tension to keep the tape straight, inserts an arrow or chalks a mark at the 20-m (or 30-m etc., as appropriate) mark of the tape. Further tape lengths are laid in a similar manner. The use of arrows is as described in § 2.2.4(a) for the laying of a chain, and working with the optical square is described in § 2.6.1.

Short tape lengths are again laid in a similar manner, but there is a common site practice of hooking the zero end of the tape over a nail in the top of a peg. If this latter method is used the peg must be very firm, and it is not recommended for accurate setting out — readings from a steel tape used in this manner, without correct tension and unadjusted for changes in temperature and possibly standardisation, should not be considered to give better than 10 mm per 30-m tape-length in accuracy.

Synthetic tapes may be cleaned with a wet cloth, but steel tapes should be wiped clean and dry before being rewound. Care must be taken to avoid twisting the tape when rewinding, and it is advisable to feed the tape through the fingers when winding in. It is important not to bend steel tapes at an acute angle. *Figure 2.6* shows how a tape is rewound in a clockwise direction.

(c) Laying the band

The band is unwound fully, in the same way as the tape, then removed from its carrying frame by lifting it off the loading peg (*Figure 2.7*). Thereafter, the procedure is as described for laying the tape, remembering that the zero mark is typically about 150 mm from the end ring. Additional care must be taken to ensure that the band is not damaged by being stood on. For accurate setting out, the spring balance should be used for each band or part band length, with the temperature estimated or recorded for each line being measured.

The correction for temperature variation (i.e., thermal expansion or contraction) may be obtained from the fact that each unit length of tape will expand or contract by 0.000 011 units per 1°C change of temperature, as in § 2.1.4. Thus a temperature correction factor f, to be applied to the field measured length, will be:

$$f = 1 + (\Delta t \times 0.000\,011)$$
$$\text{or, } f = 1 + (0.000\,011\,(t - 20))$$

where t = field temperature, 20 = standardisation temperature in degrees C, and Δt = difference

between field and standardisation temperatures. The *corrected* measurement will be:

$$f \times \text{field measured length}$$

and values of f for various field temperatures are given in the Table in *Figure 2.34*.

Steel bands are rewound in the same way as steel tapes, but using a cross-frame instead of a box.

Temp °C	Corr'n factor
- 2	0.99976
0	0.99978
2	0.99980
4	0.99982
5	0.99984
6	0.99985
8	0.99987
10	0.99989
12	0.99991
14	0.99993
15	0.99995
16	0.99996
18	0.99998
20	1.00000
22	1.00002
24	1.00004
25	1.00006
27	1.00008

Figure 2.34

2.3 Slope correction in the 'field'

Aim: *The student should be able to measure sloping distances with the chain, tape and band, both by stepping and using a clinometer.*

As stated earlier, distances are given or required in site surveys as if they had been measured on a level surface or on a horizontal plane. It will be evident that if the site is very undulating (*Figure 1.17*) and the site has to be portrayed on a flat sheet of paper, then the ground measurements up and down the slopes must be to a common datum, a horizontal plane.

Similarly, in setting out, measurements are given as if in a horizontal plane, as they have been designed on a drawing board, a flat plane.

This section deals with these problems, either by measuring horizontal distances as a series of horizontal steps or else indirectly by measuring the angles of slope of the lines being measured, then reducing the slope distances to their equivalent horizontal distances by calculation.

2.3.1 DROP OR STEP CHAINING ('STEPPING')

This method is recommended for chain surveys.

It is best carried out measuring slopes downhill, in a series of horizontal steps, using the steel tape and locating a ground point vertically below some appropriate graduation reading on the tape ('plumbing' the measurement down) as in *Figure 2.35*.

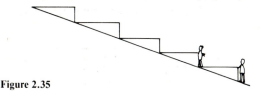

Figure 2.35

The length of the step should not exceed 20 m in chain surveys, or 10 m in accurate setting out work, and the vertical drop at the end of the step should not exceed about 1.4 m (chest height). The assistant, at the forward end of the tape step, judges whether the tape is horizontal — this may be done reasonably accurately (within 2°) provided that the tape is not above the assistant's chest height. The steel tape is normally used because the chain, being much heavier, sags and so gives incorrect horizontal distances for the steps. The ground point is 'plumbed' by dropping

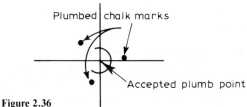

Figure 2.36

an arrow (special 'drop arrows' may be made) or a small object such as a pebble or a piece of chalk. The object should be dropped three times, to confirm the plumb point (*Figure 2.36*). The

assistant should be re-aligned before each drop of the object and, for preference, plumbing should be from a whole number of metres. The procedure is repeated for each step.

It is most important not to get confused over the number of chain or tape lengths laid if arrows are used to mark each step, and the tape should be allowed to do the adding up (*Figure 2.37*).

Although it does seem rather crude the method is relatively fast, of sufficient accuracy for chain surveys and avoids the need to correct the recorded measurements.

In setting out, a greater accuracy is often needed than in chain survey, and to achieve this the method is improved by 'plumbing down' using a ranging rod together with a small hand bubble. It is essential that the rod is not warped, and it must be vertical, hence the use of the bubble (*Figure 2.38*).

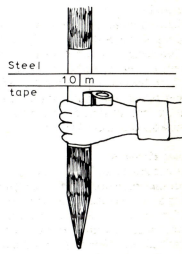

Figure 2.38 While the left hand holds rod and bubble, the right hand holds the tape sufficiently close to the bubble that both can be viewed simultaneously

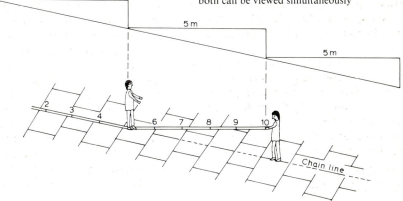

Figure 2.37 The surveyor, on left, stands at the 5-m mark lining in his assistant who, holding the tape horizontal, drops a small stone at the 10-m mark

2.3.2 MEASURING ANGLE OF SLOPE BY CLINOMETER

On long, even slopes it may be preferable to tape or chain on the surface, measure the angle of slope of the surface, then reduce the slope distance to the horizontal.

Initially, the surveyor stands close to the assistant and selects on him a mark which is at the same level as his own eye, or alternatively he may make an eye-level mark on a ranging rod. The surveyor then sends the assistant to one end of the slope, while remaining at the opposite end himself, and sights onto the selected mark with the Abney level (*Figure 2.39*). The observation

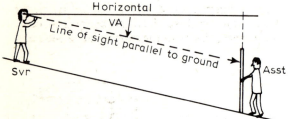

Figure 2.39 VA — vertical angle, or angle of slope

may be checked by the two individuals exchanging positions and repeating the slope measurement (*Figure 2.40*). Having obtained the slope distance and the vertical angle, the reduced horizontal distance is then calculated as detailed in § 2.4 or 2.5.

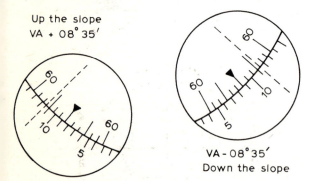

Up the slope
VA + 08° 35′

VA - 08° 35′
Down the slope

Figure 2.40

Assuming that the vertical angle is correct to within 10′ of arc when using an Abney level, and that chain survey accuracy of 1:1000 is to be achieved, then vertical angles of more than 20° should be avoided. Similarly, if an accuracy of 1:5000 is required, as perhaps in setting out, then vertical angles exceeding 4° must be avoided. An alternative is to use a theodolite, since this allows very high accuracy in measuring the vertical

angle. Yet another method is to record the difference in height between the two ends of the slope and calculate the horizontal distance from Pythagoras' theorem (*Figure 2.41*).

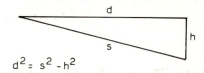

$$d^2 = s^2 - h^2$$

Figure 2.41

2.4 Slope correction — numerical solution

Aim: *The student should be able to calculate slope corrections for distances measured on an incline, using observations with an Abney level.*

Referring to *Figure 2.39* it will be seen that

Horizontal distance = slope distance × cosine of the vertical angle

$$\cos = \frac{adjacent}{hypotenuse}$$

$$= \frac{horizontal}{slope}$$

or horizontal distance

$$= slope \times \cos VA$$

The calculations may be made using cosine tables but most site surveyors should have the use of at least a pocket scientific calculator which makes calculating much easier. Note that for some calculators minutes of arc must be converted into decimals of a degree (*Figure 2.42*).

Minutes of arc	Decimals of a degree
5′	0.08
10	.17
15	.25
20	.33
25	.42
30	.50
35	.58
40	.67
45	.75
50	.83
55	.92

Figure 2.42

Example calculation:

    Measured slope distance 82.75 m
    Observed mean slope angle +06°15′(6.25°)
    then
    Horizontal distance = 82.75 × cos 6°15′
                    = 82.75 × 0.9941
                    = 82.26 m

2.5 Slope correction — graphical method

Aim: *The student should be able to explain the graphical method of correcting distances measured on an incline.*

Though this method is not greatly favoured, it is occasionally useful. It is explained by means of an example and *Figure 2.43*.

2.6 Lifting a line

Aim: *The student should be able to describe and demonstrate how to range and measure a line over a hill and through a depression.*

2.6.1 RANGING OVER A HILL

Measurement over a hill should present no problems if the procedures described earlier are used. The problem in this case lies in the ranging of the line, and the easiest way to do this is with the optical square.

    From the top of the rise, the surveyor locates a point approximately on the alignment of the line, and with the optical square he finds a point on the true alignment and marks it. It is sensible for the surveyor to check the result by locating the point twice, first with the start point of the line

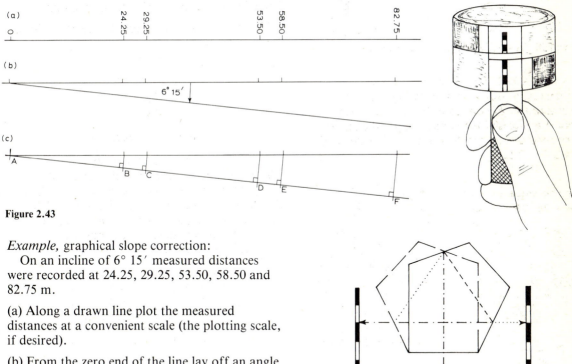

Figure 2.43

Example, graphical slope correction:
    On an incline of 6° 15′ measured distances were recorded at 24.25, 29.25, 53.50, 58.50 and 82.75 m.

(a) Along a drawn line plot the measured distances at a convenient scale (the plotting scale, if desired).

(b) From the zero end of the line lay off an angle of 6° 15′ and draw a second line through the zero point.

(c) On the line drawn in (b) erect perpendiculars to pass through the points plotted in (a), using set-square and straight-edge.
    The values of the slope distances reduced to horizontal (AB, AC, AD, AE and AF) may now be scaled off the drawing, or alternatively the method may be used directly in the chain survey plotting.

Figure 2.44

on his left then again with it on his right.
    The method of using the optical square is illustrated in *Figure 2.44*. Holding the optical square close to his eye, the surveyor walks

forward or backwards until both end points of the line appear in the field of view. When approximately on the line, the surveyor shuffles backwards or forwards, keeping the instrument and his body vertical, until he is on the true alignment (shown by the images of the two end rods forming a single vertical line in the two prisms). It is essential to avoid rocking or swaying, and with a little practice it will soon be found easy to keep the optical square vertical. When on the alignment, a mark is placed on the line of the surveyor's toe-caps, or some other identifiable point on his shoes. Turning about, he repeats the procedure while facing in the opposite direction and places a second mark which should be within 75 mm of the first. A rod is now placed at the mid-point between the marks.

If no optical square is available, an alternative method is for the surveyor and his assistant to repeatedly line each other in (*Figure 2.45*). The illustration shows the plan view of a hill, with ranging rods at the end points A and B which are not intervisible but between which a line has to be ranged. The surveyor stands at C, a point considered to be on the approximate alignment and as close as possible to A, yet with ranging rod B in view. Similarly, the assistant stands at a point D, on the assumed approximate alignment, as close as possible to B while keeping rod A in sight.

The site surveyor then directs the assistant onto line CB giving a new point D' on that line. In turn, the assistant directs the surveyor onto the line D'A, giving a new point C'. The points D' and C' are closer to the required alignment than were the original points D and C, and starting from D' and C' the process is repeated to produce points D'' and C'', and so on until neither individual can move. When this position is reached, both are located on the required alignment, the straight line between A and B.

2.6.2 MEASURING THROUGH A DEPRESSION
Ranging through a depression is not generally a problem, but if necessary methods as described in

§ 2.6.1 may be used. Similarly, measuring is usually simple, steep slopes being stepped as described in § 2.3.1, but for a short and deep depression it may be necessary to measure *in catenary* (*Figure 2.46*). In catenary measurement

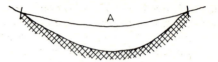

Figure 2.46 Steel tape sag, A, in catenary over a depression

the tape or band is supported at its ends while the length is allowed to hang unsupported in a free curve between its end-points, just like suspension bridge cables. (The term comes from the mathematical name of the curve naturally adopted by a chain supported only at its ends.)

Although the tape sags, no correction for sag need be applied provided that approximately the correct tension is used and that the unsupported span does not exceed 10 m (in chain surveys, 20 m) for the more accurate work. The most accurate taping methods (better than 1:10 000) make use of this technique but they are not covered in this book. Lengths greater than 10 m may be supported by intermediate supports at intervals of not more than 10 m, or alternatively the tension applied to the tape may be increased. For more details see Building Research Establishment Digest No. 234, February 1980: *Accuracy in Setting-out*.

2.7 Measuring across or over or through an obstacle
Aim: *The student should be able to describe and demonstrate how to measure around a pond, across a river or busy road, or around or through a building obstructing vision.*

The methods described here have been applied by the authors and are considered to be the best. There are many alternatives described in other

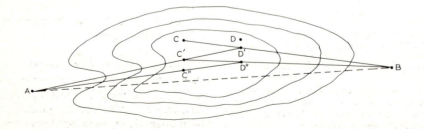

Figure 2.45

texts, but with many of these there will be difficulty in finding a site large enough to set out the required geometric shapes, and in others there will be problems in adhering to the principles of surveying. Note that it is always best to use the simplest and most direct method — if it is possible to 'throw a tape across a river' then there is no point in using one of these methods, and where possible it is always better to measure straight through a building rather than around it.

The methods outlined in §§ 2.7.1 to 2.7.4 are typically used where a straight line XABY can be ranged visually, but it is not possible to measure the section AB directly. (A and B may be located on either side of a deep river or pond, a heavily trafficked road, or some similar obstacle to direct measurement.)

In some of these methods it is necessary to lay out one line at right angles to another (*raise a perpendicular* or *raise an offset*). If low accuracy is adequate (say, correct to 50 mm) then the methods of raising an offset used in chain surveying may be appropriate, as described in § 3.1.1(b). If higher accuracy is demanded then the right angle should be set out by theodolite (see Part D) or by constructing a triangle with sides in the ratio 3:4:5 using one or two tapes.

The latter technique is illustrated in *Figure 2.47* where the zero and 27-m marks of a tape are placed and held at points 9 m apart on the chain line, the tape then being pulled taut at the 12 m

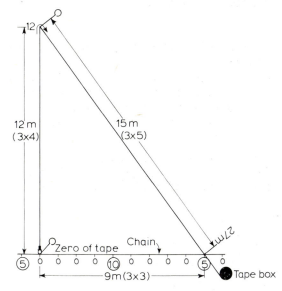

Figure 2.47 Zero and 27-m mark of the tape are positioned and held at two points 9 m apart on the chain (or other tape). Tape is pulled tight and an arrow inserted at 12-m mark, thus forming a right angle at the zero of the tape

mark and an arrow inserted. The resulting triangle has sides of 9, 12 and 15 m (ratio 3:4:5), with the 9 m side lying along the chain line and a right angle being formed at the zero mark of the tape. Any other ratios forming a right-angled triangle could be used, but 3:4:5 is easy to remember.

2.7.1 SINGLE 'A' METHOD

In *Figure 2.48* it is required to find the distance AB along the line XY. Construct a well-conditioned triangle (one in which all its angles are less than 120° and greater than 30°) ABC, such that the line DE from the mid-point of the line AC to the mid-point of the line BC is clear of obstacles. Measure DE.

Now AB = 2 × DE, since the triangles ABC and DEC are similar.

Note: DE should be measured to twice the accuracy required for the measurement of the line XY.

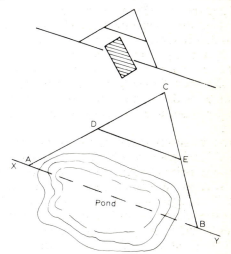

Figure 2.48

2.7.2 SINGLE 'X' METHOD

Again (*Figure 2.49*) it is required to find the distance AB along the line XY. Construct two lines, AC and BD, clear of all obstacles, such that they intersect at their mid-points E. Measure CD.

Now AB = CD, since the triangles ABE and CDE are congruent.

2.7.3 INACCESSIBLE POINT METHOD

In *Figure 2.50* it is required to find the distance AB along the line XY. Erect a perpendicular, AC, on the line XY. At the point C, construct a right angle BCD, such that D lies on the line XY.

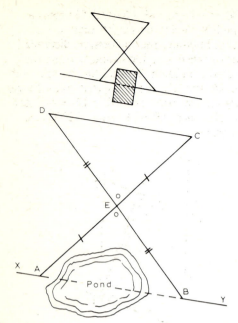

Figure 2.49

Now $AB = (AC)^2/AD$, since the triangles ABC and ACD are similar.

This method can be most useful, but it is recommended only for chain surveys. The distance AC should be as long as possible, but it need be no longer than the unknown length AB. The construction of the right angles and the measurement of the distances, for which the steel tape is used, should be to a fairly high standard of accuracy. Consider the following numerical examples:

If AC = 8.00 m, and AD = 4.00 m, then
    AB = 16.00 m,
but
if AC = 8.01 m, and AD = 3.99 m, then
    AB = 16.08m,
thus a 10-mm error in each measurement, or in the raising of the right angle, can cause an error of 80 mm.

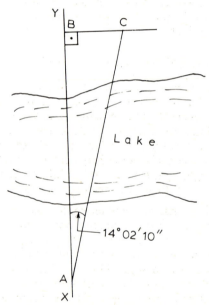

Tangent $14°02'10'' = 0.25$ (i.e. $\frac{1}{4}$)

and $\tan{}'A' = \dfrac{BC}{AB}$

$\therefore \dfrac{BC}{AB} = \dfrac{1}{4}$, or AB = 4BC

Figure 2.51

In triangle ABC, $\angle CAB = 90°$
$\therefore \angle ABC + \angle BCA = 90°$ 1

In triangle ACD, $\angle DAC = 90°$
$\therefore \angle ACD + \angle CDA = 90°$ 2

By construction, $\angle BCA + \angle ACD = 90°$ 3
$\therefore \angle ACD = \angle ABC$ from 1 and 3
and $\angle BCA = \angle CDA$ from 2 and 3

$\therefore$ the triangles are similar

$$\frac{AB}{AC} = \frac{AC}{AD}$$

or $AB = (AC)^2/AD$ **Figure 2.50**

2.7.4 THE 14.02.10 METHOD

This simple and accurate method (*Figure 2.51*) requires the use of a theodolite, but the reader who is not yet familiar with the instrument may nevertheless appreciate the method and put it into practice when he has covered Part D and has had the opportunity to handle a theodolite. The name of the method comes from the fact that a theodolite must be used to set out an angle of 14°02′10″.

It is required to find the distance AB along the line XY. At B, construct a right angle using a tape or the optical square. At A lay off (set out) an angle of 14°02′10″, such that it intersects the right angle raised from B at the point C, and measure BC.

Now AB = 4 × BC.

In practice, angles other than 14°02′10″ could be used, but the advantage of using this particular angle is that the perpendicular BC is relatively short as compared with the unknown distance, therefore errors in the setting out of this right angle will usually be minimal. The multiplication factor of 4 is also an easy number to multiply by, and to maintain the accuracy of the line XY it follows that the perpendicular BC must be measured to four times the accuracy required of the line itself, and again this is usually attainable.

2.7.5 MEASURING THROUGH A RECTANGULAR BUILDING

It is often necessary to measure through a building, but first it is usually essential to ascertain where the line actually enters and leaves the building. If this cannot be done, some re-arrangement of the lines will be needed. Ranging the line through a building may be done by (a) actually sighting through it, (b) sighting over the building, or (c) lifting the line over the building.

(a) *Sighting through the building:* Modern properties often have 'through' rooms with windows at either end. If these are suitably placed, then the line may be ranged using the optical square as described in § 2.6, but here, instead of starting at the top of a hill, the surveyor stands close to one of the windows at the outside of the building and, by moving parallel to the window, he may range the line through the two panes of glass.

(b) *Sighting over the building:* This is often possible if the building is single storey (e.g. a garage) or it lies in a hollow or valley. The surveyor stands at the start point of the line, identifies the far point, then places the assistant on line at the near face of the building. The procedure is then repeated from the far point of the line.

(c) *Lifting a line over the building:* This may be possible on flat roofed buildings with access, line points being marked on the roof just as they

would be on the ground. *Figure 2.52* shows a chain line XY passing over or through a rectangular building, the inaccessible distance being the distance AB, and C and D being the corners of the building. To find the distance AB, measure AC and BD (assuming that AC is longer than BD) then measure out BE so BE = AC, and measure CE. Then AB = CE, since ABEC is a parallelogram.

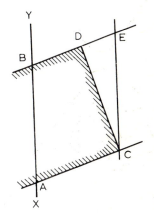

Figure 2.52

Should other detail obstruct the line DE and/or EC, then the length AB can be derived using Pythagoras' theorem (*Figure 2.53*).

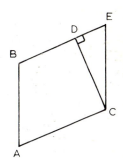

$$EC^2 = DE^2 + DC^2$$
now $EC = AB$
and $DE = BE - BD$
but $BE = AC$
∴ $DE = AC - BD$
or $AB^2 = (AC - BD)^2 + DC^2$
∴ $AB = \sqrt{(AC - BD)^2 + DC^2}$

Figure 2.53

If the offset to C from line AB is required, then this may be achieved as shown in *Figure 2.54*.

Extend AC to F such that CF = AC,
extend BD to G such that DG = BD,
then AB = FG,
the distance AO = FP, a perpendicular having
been dropped from C to FG at P,
and the offset OC = PC.

Note: These methods may need to be modified if
the wall faces are not mutually perpendicular.

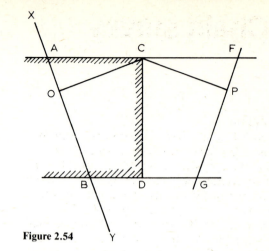

Figure 2.54

3 Chain survey

Introduction

Sections 2.1 to 2.7 refer to the measurement of lines as used in chain (offset) survey and also in building surveys, setting out, traversing, etc. The following sections deal specifically with the use of such measurements for plan production by the chain survey method, where scales of 1:500 or smaller (e.g., 1:5000) are to be used. The method is *not* recommended at scales greater than 1:500 because the accuracies usually attained in the ranging and measuring of lines are generally inadequate at the larger scales (e.g. at 1:200).

Section 1.2.1 (a) (supplying detail by rectangular offsets) should now be revised.

3.1 Offsets, straights and plus measurements

Aim: *The student should be able to measure offsets and ties by optical square and tape.*

Generally, detail in chain surveys is supplied by rectangular offsets from chain lines. In site surveys, however, it is not always feasible to use offsets and alternatives known as *ties* (or *tie lines*) are sometimes needed. In areas of 'close' detail, in fact, considerable flexibility in the methods used is desirable. The methods used by the site surveyor in practice are described here.

3.1.1 OFFSETS

An offset has been defined as a short measurement taken (raised) at right angles from the chain line to the point of detail to be surveyed (*Figures 1.1 and 1.2*). Offsets are then said to be 'raised' and 'measured'.

(a) *Recommended limitations:* Offsets to points of detail such as corners, junctions and ends of detail should not normally exceed 8 m in length for scales of 1:500, 1:1000 and 1:1250, nor should they exceed 16 m for scales of 1:2000 and 1:2500.

Similarly, offsets to curving and indefinite detail such as meandering streams and hedges should not exceed 16 m in length for the larger scales or 20 m for the smaller scales.

These limitations are imposed by the need to ensure that the survey measurements are compatible with the maximum accuracy which

may be demanded in the plotting, and the need to avoid long offsets which would require much additional care and time to maintain the possible required accuracy.

(b) *Raising and measuring an offset:* The recommended procedure for the majority of offsets to be raised and measured is as follows:

The assistant carries the synthetic tape for measuring the offsets and, on being informed by the surveyor that an offset is required at a particular point of detail, he hands the tape box to the surveyor. Unless otherwise instructed, the assistant places the zero end of the tape at ground level on the point of detail to be surveyed. The surveyor holds the tape horizontal and taut at the approximate point on the chain where the offset is to be measured (*Figure 2.9*). Up to an offset length of 8 m, the site surveyor should have no problem in judging by eye the right angle required between the tape and the chain line correct to within 0.05 m. If the surveyor gently swings the taut tape from side to side in a horizontal plane like a slow moving pendulum it may assist his judgment of the right-angle and location of the correct point on the chain line.

Offsets up to 4 m in length are often more quickly raised and measured with a 2-m ranging rod rather than the tape.

When the line of sight between the chain and the detail point is obstructed, or the offset is more than 8 m in length, it is recommended that the right angle be raised with the optical square rather than by eye.

It will be shown later that some offsets are raised but not measured. Up to a distance of 8 m the surveyor may judge the location of the offset point on the chain quite easily without tape, pole or optical square. This is done by the surveyor standing and facing the point of detail to be surveyed with his toes close to the point on the chain at which it is anticipated the offset should be raised. It will then be found that the surveyor can, with ease, place the toe of his shoe on the chain at the required point.

3.1.2 RUNNING OFFSETS

Running offsets are offsets to two or more points

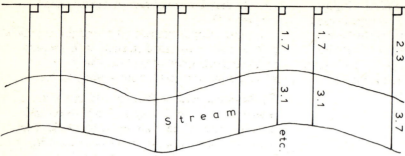

Figure 3.1

of detail along the same perpendicular, as for example to both banks of a stream (*Figure 3.1*). The limitations, and the methods of raising and measuring, are the same as for ordinary offsets. Such offsets are called 'running' because the measurements are what are known in surveying as *running measurements*, that is to say successive measurements recorded along a line from some common point. *Figure 3.2* illustrates

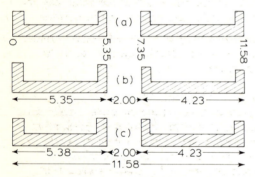

Figure 3.2 (a) Running measurements, (b) separate measurements, (c) separate and overall

such a set of measurements, as do the measurements along a chain line itself (*Figure 1.1*). These forms of measurement are widely used in survey.

With experience it will be appreciated that running measurements, including running offsets, provide the quickest and most accurate method of measuring to a number of points on the same straight line.

3.1.3 BRACED OFFSETS

Braced offsets (also known as *ties* or *tie lines*) are measurements taken from two or three different points on a chain line to a common point of detail which is to be surveyed. Unlike the ordinary perpendicular offset, braced offsets may be at any angle to the chain line.

This type of offset is used when it is considered that the length of a simple offset would be excessive, or when it is not possible to raise a right angle between the chain line and the point of detail (*Figure 3.3*). Though two measurements

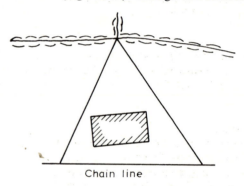

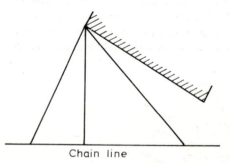

Figure 3.3

are usually sufficient over short distances, three may be preferred when long tape lengths are needed.

The term 'braced offsets' is preferred, in order to avoid possible confusion with the 'tie line' sometimes used in chain survey to denote a chain line which makes a rectangular network of chain lines plottable by tying them together to form a triangular network (line BD in *Figures 1.4(a)* and *(b)*).

(a) *Recommended limitations:* A point of detail to be fixed by braced offsets should lie within 20 m of the chain line, unless a steel tape is used for the measurements. The length of the base of the triangle formed on the chain line should be the longest side length, so that a good 'cut' may be ensured when the point of detail is plotted. (*Figure 3.4*).

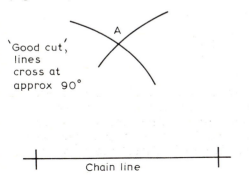

Figure 3.4

(b) *Measuring braced offsets:* Distances up to 20 m may be measured with a synthetic tape but longer lengths should be measured with a steel tape.

3.1.4 STRAIGHTS

A *straight* is the extended alignment of any straight feature, including features such as the side of a building, a straight wall, etc. The intersection of a chain line and a straight is known as the *'point of intercept'*. The straight is little used by the site surveyor in chain survey, but it may be useful from time to time in areas of 'close' detail and in the unfortunate circumstance of the surveyor not having an assistant. The straight may be 'measured' or 'unmeasured' (*Figure 3.5*).

It must be appreciated that a straight cannot be plotted from a recorded straight measurement unless the other end of the straight is also surveyed by an offset, another straight, etc.

(a) *Recommended limitations:* To minimise plotting errors which may arise due to possible misalignment of the chain lines, the angle at the point of intercept between a straight and the chain line should not be less than 40°.

Measured straights should not exceed 20 m in length, this meaning the distance along the straight between the point of intercept and the detail. Unmeasured straights may be of any length, provided that there is no doubt in the mind of the surveyor as to where the straight cuts the chain line.

It is often preferable that the chain line be arranged so that one end of a straight detail is close to the chain line, say within 20 m, but this will depend upon the length of the detail and the clarity with which the straight can be defined.

(b) *Measurement of straights:* Measured straights measured with a synthetic tape may be up to 20 m in length.

3.1.5 PLUS MEASUREMENTS

Plus measurements are measurements made at right angles to a length of straight detail, often known as an *offsetted base*, which has been surveyed by offsets, straights, etc. Plus measurements are used to enable rectangular buildings to be supplied from one or more chain lines (*Figure 3.6*).

(a) *Recommended limitations:* There should not be more than three plus measurements from

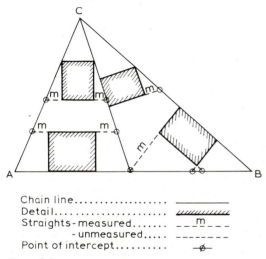

Chain line.....................	———————
Detail........................	⫻⫻⫻⫻⫻⫻⫻
Straights - measured........	– –ᵐ– –
- unmeasured.....	– – – – – –
Point of intercept..........	—∅—

Figure 3.5

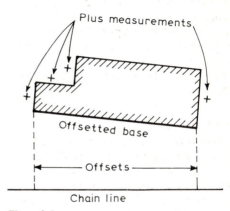

Figure 3.6

either end of an offsetting base (*Figure 3.6*, left-hand side). The total length of the sum of the plus measurements at either end of an offsetted base should not exceed the length of the base, or 20 m, whichever is the least. The recommendations are intended to ensure that the plotting shall be as accurate as possible.

(b) *Measuring of plus measurements:* Again, the synthetic tape may be used for all readings up to 20 m.

3.2 Chain survey booking

Aim: *The student should be able to record measurements taken in the field, using a recognised booking method.*

The recording of chaining measurements, known as *chain survey booking*, is carried out either on *booking sheets* or in a *field book*.

The field book can be used solely for chain survey work recording or it may be used to record all the measurements required for some particular survey task. In the latter case, the field book might contain theodolite observations, chain survey measurements, and even the level bookings recorded in carrying out the same job.

The ideal booking sheet or field book for chain survey is A4 sized and of plain or 5-mm square ruled paper. Chain survey books as obtainable from stationers or suppliers of office materials are usually small octavo size (between A5 and A6) and characterised by either a single or two red lines 15 mm apart down the centre of the page. While suited to the beginner, in practice these will be found inadequate since the page size is restrictively small and the position of the red lines may not always be convenient. With the recommended type, the surveyor may rule any lines to suit his own convenience.

Methods of booking vary but most surveyors adhere to a small number of basic rules which enable any other site surveyor to understand what has been recorded. This is useful for a number of reasons, one being that it is possible for someone else to plot the work, a desirable feature since it goes some way towards providing the independent check identified in the principles of survey. It also means that should the surveyor be unable to complete the work for any reason, then others can interpret what was intended, what has been completed and what is still outstanding. Again, the site surveyor's immediate superior will be able to assess how the work is progressing and criticise or compliment as appropriate.

The booking methods recommended by the authors are well tried on large and small jobs in rural and semi-rural areas, and in 'close' detail areas such as factories, hospitals, town centres, railway sidings, etc. Most site surveying jobs are of a simple nature, and modification of the rules described will generally be acceptable—examples of such modifications will be given later.

Clarity and accuracy of booking are obviously essential. These may be achieved by neatness, taking especial care with one's handwriting, numbering and possible emphasis of detail, i.e. printing the numbers and using a slightly broader gauge of line to represent the detail. It is preferable to book in ink, unless the weather is inclement, a fountain pen being better than a ball-point pen for forming numbers and for line work. The site surveyor must learn to do the bookings once only, since hand copying will lead to errors and considerably increase the time taken on the job, but a satisfactory presentation and speed will come with practice. Mistakes inevitably occur, but no harm is done if these are carefully corrected—incorrect figures should be cancelled by drawing a single line through them, and revised values printed above or below the original figures, while cancelled detail should be indicated by small crosses on the cancelled lines.

As will be shown later, the figures written on a page of field bookings may be written in different directions, according to the direction of measurement. Should there be any possibility of ambiguity as to the way up figures should be read then they should be underlined. For example, if 61 is likely to be confused with 19, or 8.6 with 9.8. Explanatory notes may be added to assist the plotter, and the numbering of pages is helpful and also provides a check as to whether any loose sheets may be missing.

It is unreal to expect the surveyor to be able to keep the pages of a field book clean, since chains and tapes become dirty and there is no way of preventing some of this dirt from being transferred to the book unless an extra (generally uneconomic) assistant is employed.

3.2.1 THE OPENING PAGE(S) OF THE FIELD BOOKING
This page, or pages, should contain a diagram showing the layout of the chain lines, an extremely useful feature when the survey contains many of these. The diagram provides an index to the chain lines, indicating the page on which the chainage and measurements to the detail supplied from a particular chain line may be located. It also makes plotting the lines easier, since

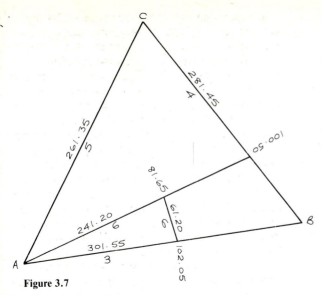

Figure 3.7

reference to all the chainage and offset values in the remainder of the book is not necessary.

A form of presentation for this diagram is illustrated in *Figure 3.7* which shows five chain lines, each represented by a solid line together with two numbers, one above the other on each side of the line. The numbers above the line, usually the larger and with their bases on the line, are those lines' measured lengths, for example, BC is 281.45 m. The numbers below the line are the page numbers in the field book on which the chainage and offset values to the detail will be found.

It will be noted that the chain lines start and finish either at the terminal points of other lines or at some point on another chain line. The terminal points of the former are known as (chain survey) *stations*, lettered A, B and C in the example diagram. The terminal point of a chain line that ties out at some point along another chain line is known as a *tie point*. The tie point values are also given in the diagram; for example 102.05 represents such a tie point value. There is no hard-and-fast rule as to which side of 102.05 the line length and page number (301.55/3 here) should be written, it is written where space best permits.

If the plotter is to plot 61.20/6, then he must know whether the tie point value of 102.05 m is the measured distance from station A or from station B. This is indicated in the diagram where the base of the figures of the tie point value is towards the start point of the line, i.e. the distance of 102.05 m is measured from station A. A check on this is provided by the mode of

writing the length of the line. Most surveyors, when writing or printing the overall length of a line on a diagram, follow the convention that the start point is to the left of the figures, thus in the case of line AB it will be observed that point A was the start point since it lies to the left of the figures 301.55. Had the line been measured from B to A, however, the surveyor would have turned the sketch around and written the figures 301.55 on the *other* side of the line. Similarly, it may be seen that line 281.45/4 was measured from C to B, line 261.35/5 was measured from A to C, and so on.

Alternative methods which may be encountered include a variation of this method in which the tie points as well as the stations are lettered, and a method in which the lines are identified by numbers with the direction of measurement of each denoted by an arrow. As the surveyor becomes more experienced, however, he may well be called upon to carry out more demanding survey jobs, when it will be found that alternative methods, which may initially appear simpler than the recommended type of diagram, do not fulfil its objectives. As stated, these are to enable the plotter to abstract at a glance the measured length and values of the tie points without having to 'wade through' the mass of chainage and offset figures in the body of the book, and also to act as an index or contents page to the chain lines.

A sketch, an outdated plan or a smaller-scale plan attached to the field book will be found useful at the reconnaissance stage, and the proposed stations and chain lines may be drawn on this. The first one or two pages of the field notes should also include relevant information such as the name of the site or the job, the address, the name of the client if appropriate, the surveyor's name, the date, etc.

3.2.2 CHAIN LINES, CHAINAGE ENTRIES, AND LINE LENGTHS

Within the main body of the field book, a chain line is usually represented by two parallel lines about 15 mm apart. The chainage figures are entered between these parallel lines, commencing at the *bottom* of the column, the distances to be entered being running measurements from the commencement of the line. The readings should be entered sequentially, and so spaced as to allow any entry to be amended or an additional entry inserted, and no attempt should be made to observe scale. In this context, note that in *Figure 3.8* the same spacing has been given to the two

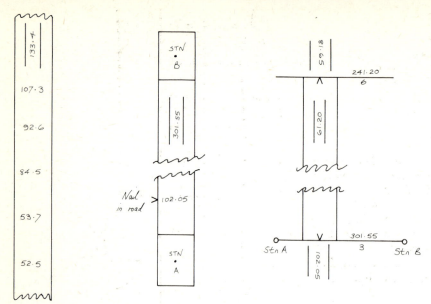

Figure 3.8   Figure 3.9   Figure 3.10

readings 52.5 and 53.7 as to the pair 92.6 and 107.3. The length of the line is entered in the column on completion of the line, and it is emphasised either by circling it or 'topping-and-tailing' it, that is, entering the length sideways on between parallel lines in the direction of chainage, as in *Figures 3.8* and *3.9*. If plain paper is used for booking, guide lines in pencil to represent the chainage column are a useful aid, and they may be overruled in ink on the completion of each chain line.

3.2.3 STATIONS, PICKETS AND TIE POINTS

A *chain survey station* is represented by a square box at the end of the chainage column, the station letter (or number) being written within the box (*Figure 3.9*).

A *picket* is a point on the ground, such as a wooden peg, nail, etc., placed in position on a chain line by the surveyor to serve as a reference mark for future use. Pickets are normally located at the possible tie points of other chain lines, being represented in the field book as shown in *Figure 3.9* which illustrates a picket consisting of a nail in the road at 102.05 m from station A on the line to station B.

A *tie point* is the start or close of a chain line either from or onto another chain line. The two tie points shown in *Figure 3.10* represent the ends of a chain line 61.20 m in length, commencing at tie point 102.05 on the line from A to B and closing at the tie point 81.65 on the line 241.20/6. It should be noted that the tie point values are

'topped and tailed' for emphasis, and that the previously-measured chain lines are repeated as single lines only, known as *skeleton lines*. Only the essential information about these skeleton lines is given, that is to say their length, page number, and value of the picket (tie point), all values being recorded and aligned as in the original chain lines. This duplication of information may not be necessary if the site is small and the plotting is to be carried out by the surveyor himself.

Occasionally, on his arrival on site to measure from or to a picket, the surveyor may find that the line originally planned is obstructed, perhaps by a heavy vehicle. If the obstruction cannot be moved then the surveyor may select an alternative chain line, thus either one or both of the planned tie points may have to be altered. Such a change may be booked as in *Figure 3.11* where 4.20 is the distance along the previous chain line between the original picket and the newly selected tie point. This measured 4.20 is generally written in the direction of measurement with the base of the

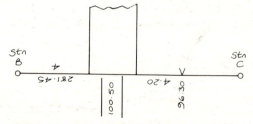

Figure 3.11

38

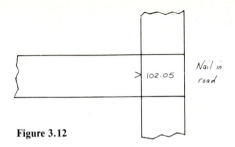

Figure 3.12

figures adjacent to the skeleton line. The value of 4.20 is shown in the booking because it provides a check for the plotter, many surveyors in practice inadvertently applying this small measured value the wrong way round (in this example, subtracting instead of adding). With the larger A4 format it may be possible to book both chain lines on the same page, thus avoiding the need for referencing between the two lines (*Figure 3.12*).

3.2.4 OFFSETS AND RUNNING OFFSETS

If a point of detail lies to the left of the chain line (i.e., on the surveyor's left as he stands on the start point and looks towards the far end of the line) then the value of the offset to the detail is written to the left of, and on the same line as, the appropriate chainage figure, but outside and as close as possible to the chainage column. The detail is then drawn freehand in a broader gauge of line (*Figure 3.13*). Where detail lies to the right of the chain line then similarly the offset value and the detail are shown to the right of the chainage column in the book.

To avoid ambiguity in linking the value and its detail, the offset value is written close to the chainage column, and the detail is drawn close to the offset value. Mistakes in interpretation are less likely if adequate spacing is allowed between the chainage values, and some surveyors find squared paper useful for showing the relative alignments.

Running offsets are also depicted in *Figure 3.13*, and this illustrates the need to enure that each offset value is shown close to the point of detail to which it refers. As an example, chainage value 103.30 m gives an offset of 2.00 m to the near feature while chainage value 109.95 m gives

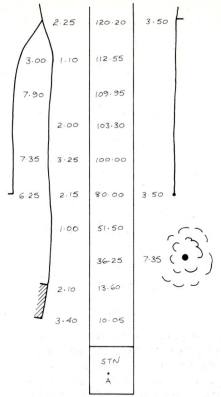

Figure 3.13

an offset of 7.90 m to the far feature. This should indicate to the plotter that the far detail is straight between the chainage values of 100 to 109.95 m and similarly the near detail is straight between the chainage values of 103.30 and 112.55 m.

All detail is assumed to be straight between consecutive offsets to the same detail, however much the freehand line drawn to represent the detail may meander, *unless* the surveyor states otherwise in the field bookings. In other words, the bookings are diagrammatic only. When taking offsets to curving detail, the usual practice is for the site surveyor to raise and measure them at such intervals that for all practical purposes the curve between the offsets is assumed to be a straight line (*Figure 3.14*).

In urban areas it is common for a number of buildings to lie on the same alignment. Where such an alignment of buildings is to be surveyed

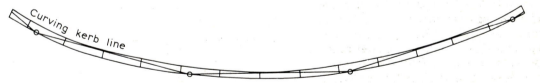

Figure 3.14 Offsets to curving detail. Spacing is dependent upon the scale of the map and radius of the curve

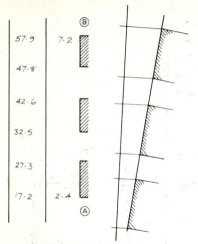

Figure 3.15

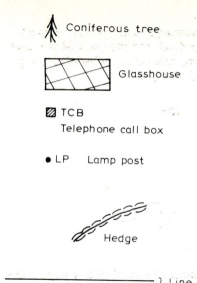

Coniferous tree

Glasshouse

TCB
Telephone call box

● LP   Lamp post

Hedge

by chaining, then it is not necessary to raise and measure offsets to every building corner on the alignment, as this practice is time-consuming and may lead to poor representation of the detail on the plotted plan. Every survey measure and every plotted point will have some accidental error, so that if such a row of buildings is each surveyed separately then, while each building should be correct on the plotted plan, the final alignment as a whole will not look precisely correct. This may be overcome by raising and measuring the corners at the ends of the alignment but only *raising* the others (*Figure 3.15*). Note that the letters A and B circled on the example bookings indicate to the plotter that the detail is straight between the offsets at A and B.

3.2.5 THE REPRESENTATION OF DETAIL
To the plotter, the field bookings are often the only indication he has of the site, thus not only the figures but also the representation of detail must be correct. Any symbols used must be understood by the plotter, and generally the symbol he draws on the map or plan is simply a neater representation of that used by the surveyor.

Plotting is referred to in § 3.7 and in *Site Surveying 3,* but at this stage the reader should be aware that symbols may be classified into three types — point symbols, line symbols, and area symbols — all of which are used in chain survey booking and plotting. In *Figure 3.13* all three have been used, a point symbol for the deciduous tree, a line symbol for a fence, hedge or wall, and an area symbol for a roofed area, in this case 'fringed hatching'. Other examples are shown in

} Line
symbols

Figure 3.16

Figure 3.16. The choice of symbol to use will depend upon a range of factors, including the proposed scale of the map or plan, its purpose and who is going to use it and where.

3.2.6 BRACED OFFSETS (TIES OR TIE LINES)
Braced offsets may be booked as pecked lines with the measured lengths placed within the pecked lines, e.g., ----8.40----(*Figure 3.17*).

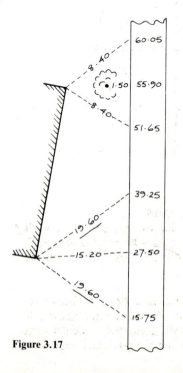

Figure 3.17

3.2.7 STRAIGHTS

An *unmeasured straight* is booked as a pecked line between the detail and the point of intercept, with the chainage value at the point of intercept circled (*Figure 3.18(a)*).

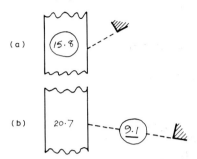

Figure 3.18

A *measured straight* is booked as a pecked line, with its measured length circled and placed within the pecked line, the chainage value at the point of intercept being entered as for an offset (*Figure 3.18(b)*).

These arrangements of circles are used to avoid any possible confusion between offsets, braced offsets and straights, since in practice it is possible for a single chainage value to refer to all three (*Figure 3.19*).

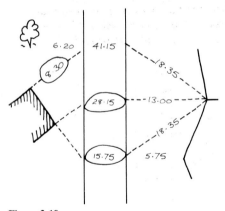

Figure 3.19

3.2.8 PLUS MEASUREMENTS

These may be booked as in *Figure 3.20*. Note that the plus sign should always be entered nearest the point of detail at which the plus measurement is taken.

3.2.9 REFERENCING DETAIL BETWEEN CHAIN LINES

In chain survey it will often be found that detail connects between chain lines, and if the surveyor is not going to carry out the plotting then it is

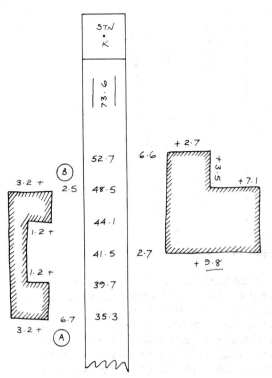

Figure 3.20

essential, particularly in larger surveys, that this detail be cross-referenced.

Where a point of detail supplied from one chain line connects through to a point of detail supplied from a later chain line, then the former point may be indicated by adding an arrowhead to the line of detail in the field bookings (*Figure 3.21*). The latter point is cross-referenced by the use of a skeleton line giving its length and page number, and repeating the former point's chainage and offset values. (*Figure 3.22*).

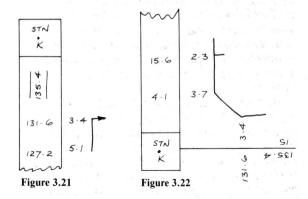

Figure 3.21   **Figure 3.22**

3.2.10 THE CONTINUATION OF A CHAIN LINE

The booking of a long chain line may require more than one chainage column, and this will

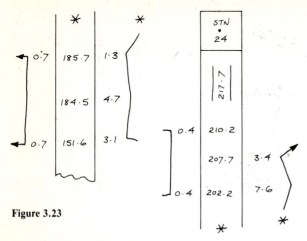

Figure 3.23

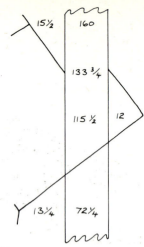

Figure 3.25 Note how the detail is shown when cutting the chain line obliquely: (i) when the chainage value is recorded, as at 133¾, and (ii) when not recorded

require cross-referencing to indicate to the plotter where the chainage column continues in the field bookings and whether there is any detail connected between the columns. This referencing may be by asterisks (*Figure 3.23*) but differently-shaped asterisks should be used for each case if there is any possibility of doubt as to which line joins which or which item of detail is continued.

The use of asterisks is also illustrated in the example field book in § 3.6.

3.2.11 CHAIN LINE RUNNING ALONGSIDE DETAIL
To indicate where a chain line runs along the line of a kerb, or the face of a building, wall, fence, etc., the line representing the detail should be drawn in a much broader gauge of line where it is contiguous with the chain line (*Figure 3.24*).

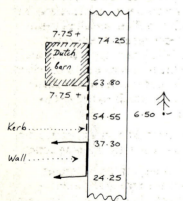

Figure 3.24

3.2.12 DETAIL 'CUTTING' THE CHAIN LINE
When a chain line crosses the detail, the chainage value of the 'cut' should be recorded only if there is a bend, corner or junction in the detail at this particular point. The reason for this is that any

small errors in alignment or measurement may cause a bend to appear in a straight feature when it is plotted. Examples of the booking of detail cutting the chain line are shown in *Figure 3.25*.

The preceding sections have covered the various techniques to be used in chain survey booking. Section 3.6. includes a complete example of the booking for a small survey task, demonstrating how these methods are combined in practice.

3.3 Chain survey equipment
Aim: *The student should be able to describe the equipment used for chain surveys.*

The following items, all described in § 2.1, form the recommended minimum set of equipment for chain surveys:

Metric chain — 20 m
Steel tape — 20 m
PVC-coated glass fibre (synthetic) tape — 10 m
Arrows — set of 10
Ranging rod — 2 m, graduated in 0.2-m bands
Optical square
Abney level

In addition to these items, the following should be available:

Prismatic compass: A small hand-held instrument for obtaining the bearing of one of the chain lines. This bearing may be used to orientate the survey. In the United Kingdom, however, the use of such an instrument is seldom necessary, since Ordnance Survey maps of the area of survey are

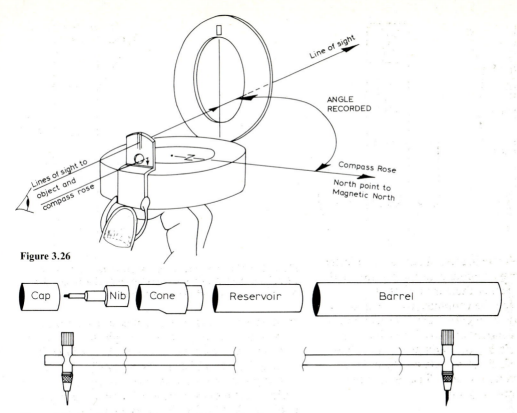

Figure 3.26

Cap — Nib — Cone — Reservoir — Barrel

Figure 3.27 A technical drawing pen and beam compass

generally available. The direction of north can usually be transferred from the Ordnance Survey map to the plotted plan with sufficient accuracy. Many varieties of prismatic compass are in use, a typical example being illustrated in *Figure 3.26.*

Ground marking equipment: Including hammer, chisel, nails, wooden pegs, grease (wax) crayon, chalk.

Booking equipment: Field book, or booking sheets with A4 clip board, pen, ink, pencil, straightedge.

Plotting equipment: Drawing material, compasses, scales, straightedge, set squares, pencils, pens, inks, erasers, etc. (a technical drawing pen and beam compass are shown in *Figure 3.27*).

3.4 Fieldwork errors

Aim: *The student should be able to identify and correct fieldwork errors.*

It should be evident that errors in chain survey fieldwork arise either in setting out alignments or in measuring distances, and that the unacceptable errors are due either to carelessness or to failure to make due allowance for the effect of cumulative or systematic errors.

As indicated in § 1.5, the methods used for measuring, booking and plotting should be designed to detect errors due to carelessness. It is not normally possible to eliminate all of these, but the use of the recommended techniques should ensure that few, if any, gross errors are not located before the plan is passed to the client. The following list identifies common blunders (gross errors) and the methods to be used to locate or avoid them:

(a) Forgetting how many 20-m chain lengths have been laid—use arrows correctly, keeping count.

(b) Booking errors, transposition of figures, misreading of tallies and tape graduations — adopt a rigid set of rules in booking, let the tape or chain do the 'adding up' when a line is broken due to a wall or step (*Figure 3.28*). The assistant should be encouraged to check tape and chain readings, especially the length of the line.

(c) Misalignment — generally due to failure to trace or range the line. If in any doubt, alignments should be checked with the optical

43

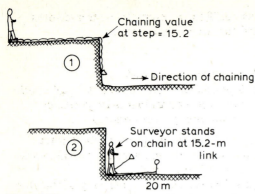

Figure 3.28

square (i.e., the independent check).

(d) Offsetted points badly positioned on plan — keep the offsets short to ensure accurate raising of the right angle and of the measured distance.

The following list identifies the cumulative, systematic or constant errors (referred to in § 1.5) which may occur in chain survey, together with the methods to be used to eliminate or reduce them:

(a) Incorrect length of tape or chain — standardise and correct as necessary.

(b) Measurement on sloping ground — use the 'stepping' method, or use the Abney level and calculate corrections.

(c) Chain sagging on undulating ground — use the steel tape, with sufficient tension (but not excessive) to remove the effect of sag.

(d) Tape sagging due to vegetation — use the chain.

If all the precautions listed have been taken and suitable methods used, yet the survey still cannot be plotted to the required standard of accuracy, then the only possible course of action is to return to the field and re-measure or check measurements, as necessary.

3.5 Selection and positioning of chain lines

Aim: *The student should be able to identify those factors which govern chain survey frameworks.*

When the surveyor arrives on site, his first task is to decide where the chain lines are to be positioned in order to survey the required detail in such a way that the whole area may be plotted

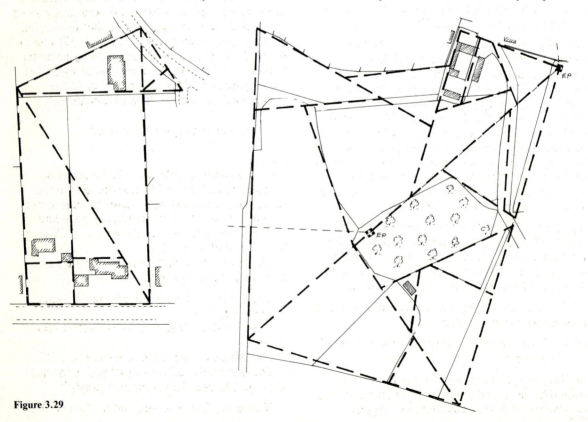

Figure 3.29

44

to an acceptable standard of accuracy, with judicious expenditure of time, effort and money. The time spent on carrying out a reconnaissance is not wasted, and it may well be the most enjoyable part of the job. The more thoroughly the reconnaissance is carried out then the easier the actual measurement of the lines will be. It must be remembered, however, that except on the very simplest of sites, it is unlikely that any two surveyors would select exactly the same arrangement of lines.

The following are the essential requirements of a good chain survey framework:

(a) The network of lines must be primarily triangular. (A simple triangle around the site would be ideal, but this is rarely practicable.) Generally there should be a base line through the site on which to construct the triangles. See § 1.2.3 regarding the need for a base as control.

(b) The triangles formed should be as few in number as possible, and they should be well conditioned (i.e., as judged by eye no angle of a triangle should appear to be less than 30°).

(c) Each triangle must be checked, therefore each triangle must be provided with one or more check lines. Again, any angles formed in the triangles should not appear less than 30°.

(d) Measurements to the detail must be kept short, therefore the chain lines should be positioned close to the detail but not so close as to cause difficulty in measuring the lines. The need to keep the chain line straight should have priority, however, and it is preferable to locate the lines along straight features or along points on the same alignment. For example, the faces of buildings or the tangent points of telegraph poles or the lines of straight kerbs or pavement edges are all suitable, provided that the resulting offset values do not become excessive.

(e) All detail which must be surveyed should be supplied by the methods described in § 3.1. It is worth noting here that many inexperienced site surveyors fail to position the lines to allow the detail at the corners of the site to be offsetted.

(f) Obstacles to ranging and measuring should be avoided wherever possible.

(g) Lines should be positioned over level ground, if at all possible.

(h) The total chainage involved in the survey should be close to the minimum required to achieve the features described in (a) to (g).

Figure 3.29 shows examples of frameworks selected for two different sites.

3.6 A practical survey task
Aim: *The student should be able to apply chain survey principles to a small practical situation (say, five stations).*

Figure 3.30 shows the pages of field notes made in a survey of eight hectares of land, to be drawn at a scale of 1:1000.

Figure 3.31 is a reduced scale copy of the plan produced from the field notes.

3.7 Chain survey plotting
Aim: *The student should be able to plot survey lines, including all detail, in accordance with BS 1192.*

Chain survey *plotting* requires no great skill, although the draughtsman must be accurate and meticulous. Graphic communication skills are of importance in the presentation of the finished work, however, and these are considered here.

BS 1192 embodies recommendations for building drawing practice, and in general these recommendations should be followed, but it must be appreciated that the object of the survey plan or map being produced may not be applicable to BS 1192. In these circumstances it may be better to be guided by a presentation similar to that of the large-scale Ordnance Survey maps (*Site Surveying 3* gives more detailed guidance).

3.7.1 PRELIMINARY CONSIDERATIONS

(a) Scale
Since the plotting scale controls the limits of measurement, it should be decided before the survey commences at what scale the plan or map is to be plotted. That is, are the chainage and offset values to be read to the nearest 0.25 m, or 0.1 m, or 0.05 m, etc.?

BS 1192 identifies the following factors as governing the choice of scale:

The need to communicate adequately and accurately,
The need to achieve economy of effort and time,
The character and size of the subject, and
The desirability of keeping all sheets for one project to one size, as far as is possible.

If accuracy is a criterion, and it often is, then

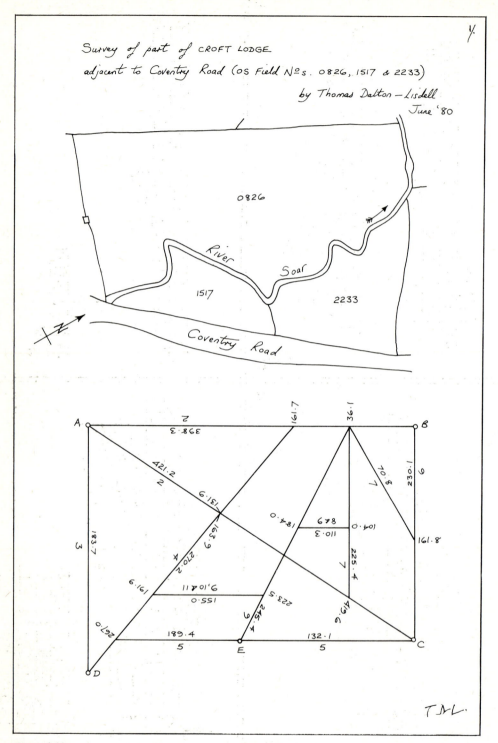

Figure 3.30

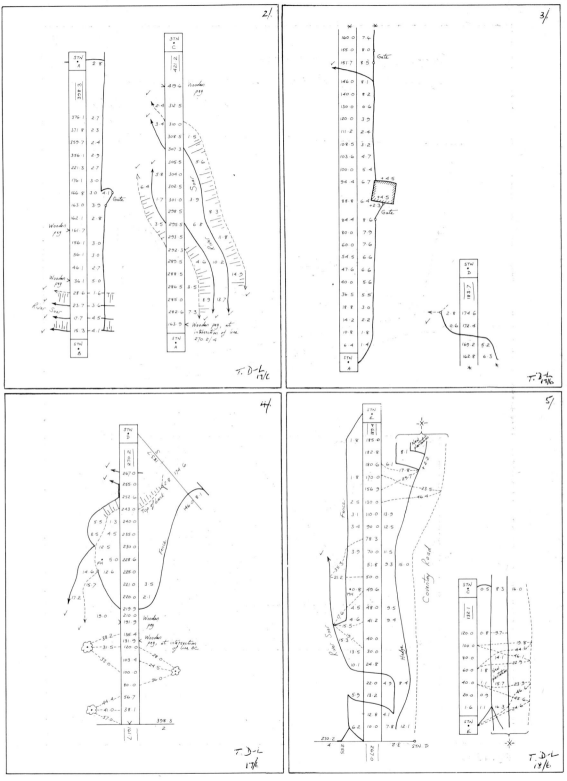

Figure 3.30 (continued)

47

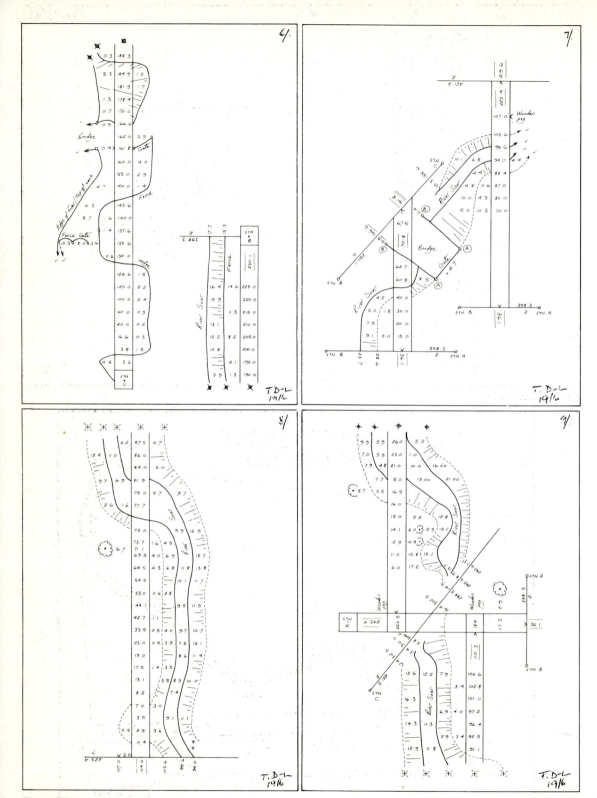

Figure 3.30 (continued)

48

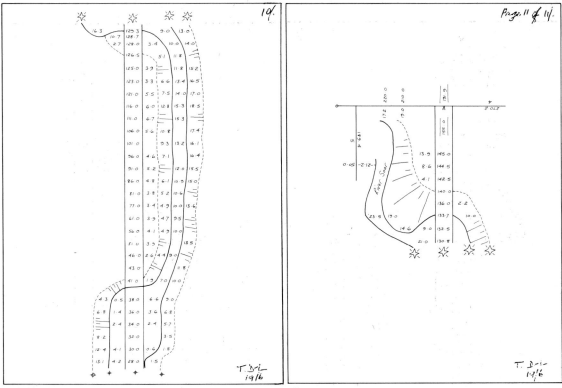

Figure 3.30 (continued)

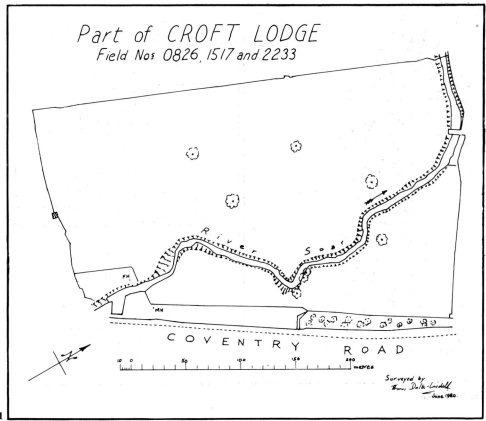

Figure 3.31

the following limits of measurement are recommended:

Scale	Chainage and measurement to detail to nearest
1:2500	0.25 m
1:1250	0.1
1:1000	0.1
1:500	0.05

Note the similarity of numerals which makes it easy to remember these figures.

The limits of measurement recommendations are based on the fact that it is possible to attempt to plot to 0.1 mm on the drawing material, therefore it should be correct to within 0.2 mm. Ground measurements, therefore, need not be better than one can attempt to plot — for example 0.1 mm on paper at 1:2500 scale represents 2500×0.0001 m on the ground, or 0.25 m, and similarly at the other scales.

(b) Sheet sizes
Sheet size is obviously linked to the chosen scale and the area to be surveyed, and on a large survey it may be desirable to have two or more sheets. If copies are required then the reprographic facilities available must be considered. Sheet size will also depend upon the amount of marginal and border information required to be shown, for example a title, a north point, a legend, road destinations, etc. BS 1192 requires that all drawing sheets should be A sizes, i.e., A0, A1, A2, A3 or A4, as illustrated in *Figure 3.32*. As each sheet is sized either half or

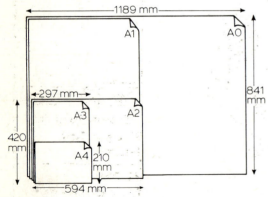

Figure 3.32

twice the size of the next larger or smaller sheet, their reduction and enlargement are simplified, and this standardisation is an aid to economical use of drawing paper and film and their easy storage. A0 = 1m².

(c) Drawing material
The choice of material for the plan or map will depend upon the purpose and use of the plotted survey plan. The material may need to be durable, to be transparent or translucent for purposes of copying and tracing, and it may need dimensional stability. The need to take ink and pencil is evident, and on occasion it may be necessary for the material to take colour washes. Also, the ability to withstand frequent erasures is generally desirable.

There are three materials in common use — cartridge paper, tracing paper and polyester-based translucent films.

If *cartridge paper* is used, then the heavier and higher quality grades are the more suitable, but a sample should be tested for ability to take ink and colour washes and to withstand erasing. Cartridge paper is not dimensionally stable, and perfect erasure of inks is not possible.

Tracing paper is probably the most commonly-used material, and again the heavier grades are better, but it is not dimensionally stable and becomes brittle with age.

Polyester film is strong, durable and dimensionally stable, probably the best material to use but it is more expensive than either cartridge or tracing paper. There is a slight tendency for ink to spread on film, and pencils must either be of a hard grade or have the more recently introduced plastic-based leads. Both ink and pencil can be erased.

(d) Layout of the survey on the drawing material
Having decided on the scale, the material and the sheet size to be used, there are still several other matters to be considered before plotting can be commenced. These include the orientation and position of the survey on the sheet, the title panel and the border and marginal information to be provided. When plotted and penned the drawing should present a balanced appearance pleasing to the eye, and much of this will be achieved by the quality of the line work and the lettering. Some appreciation is necessary, however, of the layout of the various 'bits and pieces' which make up the whole.

Orientation: On small surveys it is common practice to arrange that the principal approach (road) to the site is approximately parallel to the lower edge of the paper or film. If the area is large, it is preferable for the survey to be drawn with north facing the top of the drawing material. On occasion, other arrangements may be appropriate (see *Site Surveying 3*).

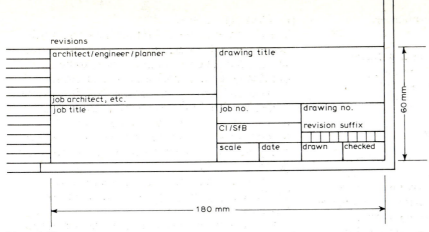

revisions				
	architect/engineer/planner	drawing title		
	job architect, etc.			
	job title	job no.	drawing no.	
		CI /SfB	revision suffix	
		scale	date	drawn	checked

Figure 3.33

Title panel: BS 1192 suggests that in order to facilitate reference when prints are filed, all drawings should have the title panel placed at the bottom right hand corner of the sheet, and this is generally worthwhile. *Figure 3.33* illustrates an example BS 1192 recommendation in this respect. The plotter should, however, be able to depart from this arrangement if it is thought necessary in the interests of accuracy of representation. Site surveys are carried out primarily to provide an accurate representation of the detail on site, and this accuracy can be lost if the plan is folded. Since the *original* should not be folded, some different filing arrangement may be needed and hence the title in the bottom right hand corner may not be suitable.

Other marginal and border information: If the

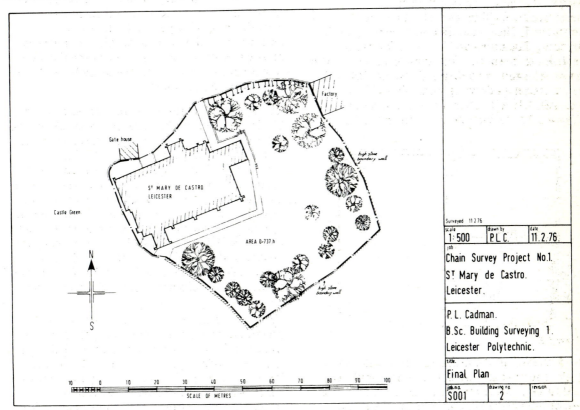

Figure 3.34

plotter intends to conform with BS 1192, then the majority of the other marginal information (scale, date, client, surveyor, etc.) will appear in the title panel. Though this tends to make it more difficult to achieve the 'balanced' look, the remaining marginal and border information should be located in the best way to try and achieve this.

Example surveys are shown in *Figures 3.31* and *3.34*, both at reduced scales.

3.7.2 PLOTTING THE FRAMEWORK — LINE PLOTTING

The equipment required includes a scale, steel straight edge (metric if graduated), pair of compasses, beam compass (*Figure 3.27*), pencil, pencil sharpener, and eraser. The pencil should be one which will produce a fine clean line, of even width and density, which will not smudge but may be erased without undue damage to the drawing surface. This means a good quality pencil sharpened to a fine point.

Plotting should commence with the drawing of the base, or longest line, in a suitable position on the drawing material and to the appropriate scale, together with any relevant tie points. Stations and tie points may be lightly encircled, with their values written adjacent to the line (*Figure 3.35*). It should be remembered that all this may have to be erased later.

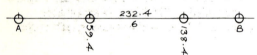

Figure 3.35

The largest and best-conditioned triangle tied to the base should be plotted next by using the compasses to describe intersecting scaled arcs of the recorded distances (*Figure 3.36*). The chain lines are then drawn in with the straight edge and their lengths checked with the scale. It is most important to ensure that the drawn lines pass exactly through the plotted points — beginners often have initial difficulty in doing this. Next, plot and check any lengths within the plotted triangle and add the relevant chain lengths and tie point values abstracted from the diagram showing the layout of the lines. The remaining chain lines are plotted in a similar manner.

For checking the lengths of lines, the limiting tolerance usually accepted is 1:500, though occasionally 1:1000 is required. Further, no line should be rejected if the difference between the

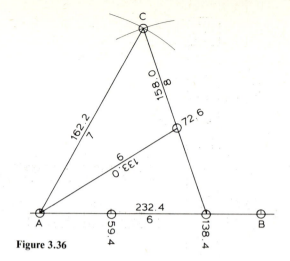

Figure 3.36

measured and the plotted lengths is less than three times the recommended limits of measurement for the scale of plotting. Thus, no lines with the following errors should be rejected:

> 0.15 m at scale 1:500
> 0.30 m at scale 1:1000
> 0.75 m at scale 1:2500

Should a line fail to plot within these limits, and the misclosure is large, then the line in error and/or its tie points must be checked and re-measured as necessary. On the other hand, if the misclosure in a line is small, then a judicious adjustment of the surrounding lines may enable it to be fitted into the plot within the above limits of tolerance. That is to say, the network holding the line in error may be expanded or contracted so as to bring all the lines within the limits of tolerance.

Where a line has an acceptable misclosure, then the plotted value of any tie point on that line is to be equated proportionally.

3.7.3 PLOTTING THE DETAIL

For this task, a 160-mm 60° set square is required in addition to the equipment listed in § 3.7.2. It is generally more convenient to plot the detail line by line, in the same order as it was chained, although this is not essential. *Figure 3.37* shows a chain line and the plot of the chainage values.

The offsets are raised using the set square and the straight edge. *Figure 3.38* illustrates the use of this equipment.

The following points in *Figures 3.37* and *3.38* should be noted:

Figure 3.37 — The chainage values have been plotted as short ticks, with

52

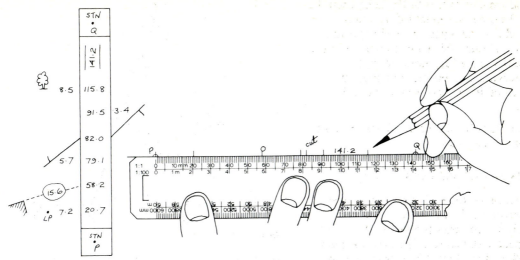

Figure 3.37

annotations or symbols to assist identification when raising the offsets and plotting the detail.

Figure 3.38 — The edge of the set square can be allowed to slide along the edge of the straight edge. The offsets are drawn to the left and/or the right, as indicated in the bookings in *Figure 3.37.*

The scale is used to plot the offset values (*Figure 3.39*) and a pair of compasses is used to plot the braced offsets or ties. The straights, whether measured or unmeasured, cannot be plotted until the far end of the straight has been

located on some other chain line.

Detail points are then connected as indicated in the field bookings, using the straight edge. Plus measurements may then be plotted using the set square, straight edge and scale in a manner similar to the plotting of offsets.

Where a line has an acceptable misclosure, the chainage values as recorded should be increased or decreased proportionally. For example, in carrying out a survey to a scale of 1:500 (plotting scale intended) the measured length of a line was recorded as 216.40 m, but on the plot the length of the line scaled as 216.80m, an acceptable misclosure of 0.40 m. If the scaled length of the

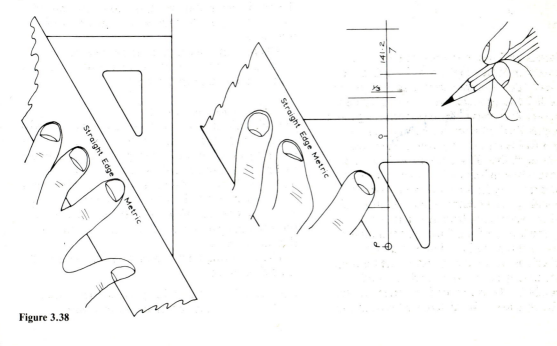

Figure 3.38

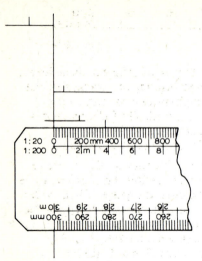

Figure 3.39

line is to be accepted, then all intermediate chainages along the line must be proportionally adjusted. The recorded lengths must be increased by +0.4 m per 216.4 m, or by +0.1 m per 54.1 m, or by +0.05 m per 27.05 m, using simple proportion. Assuming a set of recorded chainage values of:

52.05, 77.00, 99.55, 116.00, 158.25 and 185.50,
then these would have to be increased by:

+0.10, +0.15, +0.20, +0.30 and +0.35,
respectively, giving revised values for plotting as:

52.15, 77.15, 116.20, 158.55 and 185.85.

When the acceptable misclosure is at the minimum value, such as 0.15 m at 1:500 scale, 0.3 m at 1:1000 scale, etc, then the misclosure may be 'lost' at the terminal points of the line without adjusting intermediate chainages.

3.7.4 PENNING IN—COMPLETING THE PLAN
The plan or map should adequately and

accurately communicate its contents to the user, this being done by the use of line, point and area symbols (as referred to above) together with appropriate lettering and numbering. The symbols and lettering are used not merely to represent and describe the detail but also to provide all the other information needed to interpret the plan or map.

(a) Contents
The *contents* of the map or plan may include the following items in addition to the measured and plotted detail:

Names: Names describing detail, both distinctive, such as 'High Street', and those which are descriptive, such as 'RWP' representing a rain-water pipe.

Border: This is not essential, but a border often helps to give balance to the final presentation drawing. Borders should be simple, neatly drawn, and may be of one, two or three ruled lines. Note that an ample margin must be left at the left-hand edge if the plan is to be bound (for filing). BS 1192 suggests a minimum width of 20 mm.

Title, title panel: See *Figures 3.33, 3.34* and *3.40* for examples of these.

Scale and scale line: All survey plans should have some indication of their scale, such as a representative fraction (RF), for example 1:500. This is usually located in the title panel, as recommended in BS 1192, or underneath or adjacent to the title or to the scale line. The scale line, which should be kept simple (*Figure 3.41*), may be used for scaling from the original or a photographic reduction or enlargement of the map or plan, and also to give some indication of the stretch or shrinkage of the drawing material.

Survey of field adjacent
The Croft Leicester Lane Desford

SCALE   : 1 1000
SURVEYED BY : R D Matts
DATE   : 7th December 1975

CLEARDENE   FARM

Figure 3.40

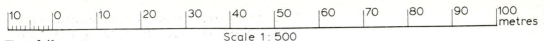

Figure 3.41

Scale 1 : 500

The minimum plan length of a scale line is usually 0.2 m, but this depends upon the sheet size and the use of the line to make the 'whole' pleasing to the eye. The scale line is generally drawn adjacent to the lower edge of the sheet, but again, to obtain balance, it may be placed elsewhere.

Names, date, job and drawing number, etc.: This information is usually placed in the lower right-hand corner of the sheet, hence it may be included in the title panel as recommended in BS 1192. This should include, possibly, the name of the client, how surveyed, the surveyors responsible, date of survey and any relevant reference numbering. (*Figures 3.34* and *3.42*).

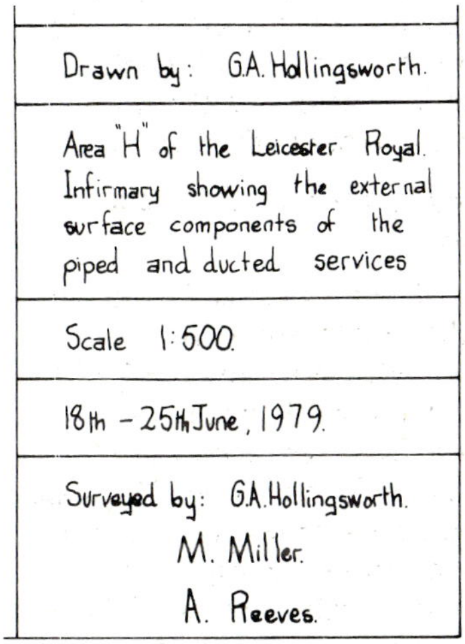

Drawn by : G.A. Hollingsworth.

Area "H" of the Leicester Royal. Infirmary showing the external surface components of the piped and ducted services

Scale 1:500.

18th – 25th June, 1979.

Surveyed by: G.A. Hollingsworth.
M. Miller.
A. Reeves.

Figure 3.42

North point: A plain north point should be placed in a convenient position, correctly orientated, to assist in making the plan or map pleasing to the eye. BS 1192 recommends the symbol illustrated in *Figure 3.43*.

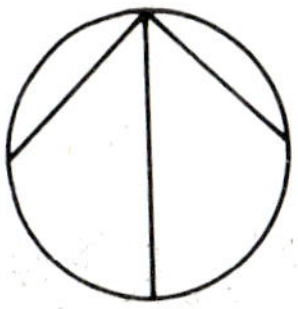

Figure 3.43

(b) Presentation
The mode of *presentation* of the map or plan has a significant effect on the way in which it communicates information to the user. This is concerned with the use of different visual levels, using symbols and lettering of varying size and form, to lend emphasis and contrast to the features being portrayed. The following notes provide some guidance as to improved presentation of the map or plan contents:

Drawing lines: BS 1192 suggests that to make drawings as clear as possible three different line widths should be used, the gauge of line for any specific purpose being constant throughout the drawing. Further, it is suggested that the ratios of the three widths should be 1:2:4, possibly 0.2 mm, 0.4 mm and 0.8 mm. Since the publication of the Standard, however, the traditional pen sizes have been superseded by the widths recommended by the International Organisation for Standardisation (ISO). The new line widths are 0.13, 0.18, 0.25, 0.35, 0.5, 0.7, 1.0, 1.4 and 2.0 mm, each size being $\sqrt{2}$ times the preceding size in the list, the same relationship as used for the international A-size papers.

As an example of these related sizes, if a sheet of A2 paper which has a line drawn on it with a 0.5 pen is photographically reduced to A3, further lines drawn on the A3 with a 0.35 pen will be of the same apparent thickness as the reduced original line.

The line widths most nearly meeting the BS recommendations are 0.18, 0.35 and 0.7 mm, or 0.25, 0.5 and 1.0 mm. On survey plans, for the beginner, two line widths are enough, and again it should be observed that 0.18 mm is not a recommended size for the inexperienced to use, while 1.0 mm is rather wide for anything except a border line. *Figure 3.44(a)* illustrates a variety of lines, and *Figure 3.44(b)* the widths.

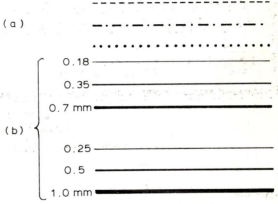

(a)

(b)

0.18
0.35
0.7 mm

0.25
0.5
1.0 mm

Figure 3.44

Drawing pens: The best pen for line drawing is the modern technical pen (*Figure 3.27*). Based on the fountain-pen principle, these pens give a consistent line width and they are easy to use if the maker's instructions are followed. Interchangeable nibs or separate pens are available for all the line widths referred to above.

Lettering: Like the line symbols, lettering may vary in form and size, provided that it is simple and neat (the essentials of legibility) and that it is unaffected. No serifs should be used, the width of letters should be constant, and the line width of the letters should be the same as used for the line drawing. Freehand lettering and figures should be well-formed and uniform, a problem for the beginner. Stencils and transfers help to achieve uniformity, particularly for headings and titles, but they are slow in use and lack personal character.

The BS recommends that the height of letters and numbers in titles should be from 5 to 8 mm, while other notation should be between 1.5 and 4 mm. Examples of acceptable freehand (a) and (b), stencilled (c) and dry transfer lettering (d) and (e) are shown in *Figure 3.45*.

Good lettering can only be achieved with a great deal of practice, and beginners may find 'sloping Egyptian' (*Figure 3.45(a)*) the easiest freehand style to use, giving acceptable results quite rapidly. Defects in lettering are more evident in upright than in sloping styles, and in larger sizes. Guidelines, both horizontal and sloping, should be ruled in pencil for practising freehand lettering, and guidelines should be ruled lightly before lettering on the plotted survey (*Figure 3.46*).

Letters should be evenly spaced within words, the words being compact without being cramped, and sufficiently close to one another to allow sentences (and groups of words) to be read without difficulty. The clear space between letters

Figure 3.46

should not be less than twice their line thickness, and the space between words should equal that occupied by the letter O touching both words. The technical pen may be used for freehand lettering.

Names: In positioning the name of a feature on a map or plan there should be no possibility of confusion as to which feature the name refers to, nor should the name obscure any other part of the drawing. These aims will be assisted if names are normally written with the base of the word(s) towards the lower edge of the map or plan, or, if a name runs vertically, its base is towards the right-hand sheet edge.

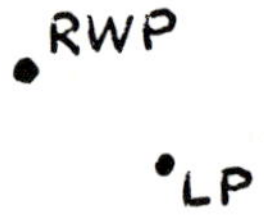

Figure 3.47

Figure 3.48

Names which identify point symbols should be placed so that the initial letter of the name is close to the symbol and preferably to the right and slightly above or below the centre of the symbol (*Figure 3.47*).

Names identifying linear features such as roads and rivers should be aligned within or alongside the feature, spaced so as to indicate the feature's extent (*Figure 3.48*).

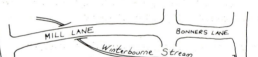

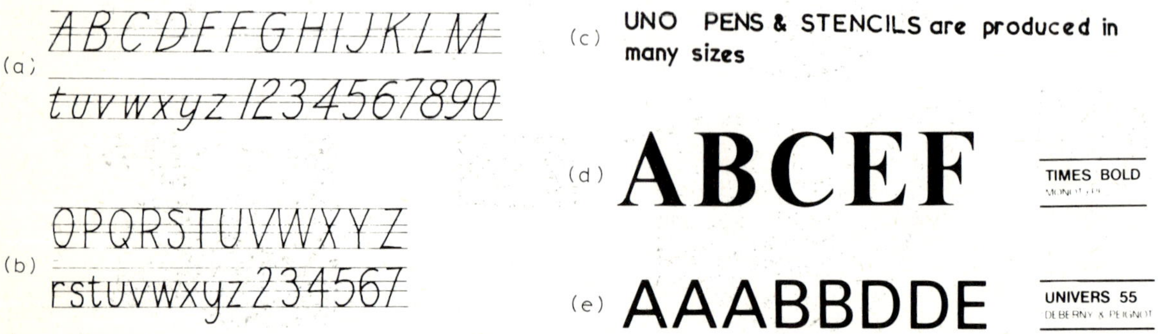

Figure 3.45 (a) Sloping Egyptian, and (b) upright, are drawn freehand

56

Names identifying areas, such as field and wood names, should be placed centrally within the area and with sufficient spread to give an indication of the extent of the area. The greater the spread of the lettering, of course, the more difficult it is to read.

Point symbols: These may vary in shape and size, and they may be freehand, stencilled, or dry transferred. *Figure 3.49* shows examples from BS

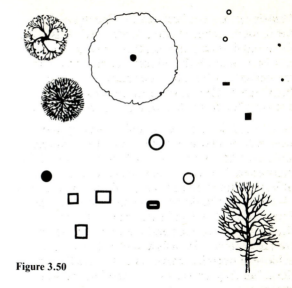

Figure 3.50

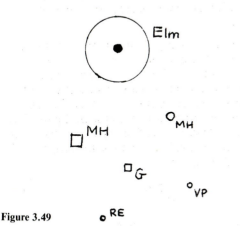

Figure 3.49

1192, but see also *Figure 3.47*. *Figure 3.50* shows examples from dry transfer sheets.

Area symbols: These are sometimes used to define roofed areas, areas of differing surface vegetation and areas of natural and man-made surfaces. Example of some of the area symbols in use are shown in *Figure 3.51*. These include a few of the dry transfer hatchings, shadings and textures which are tending to replace some of the traditional freehand methods, particularly colour washes. Colour-wash is not satisfactory if copies of a survey map or plan are to be produced by the cheaper reprographic methods.

Some of these symbols are rather decorative, and a small amount of this 'ornamentation' does enhance the finished work.

The penning-in of detail can be carried out as plotting proceeds, possibly line by line but preferably on completion of the pencil work. It must be remembered that on completion, the presentation should be such that nothing on the plan or map appears to be too small or too large, too long or too short, too thick or too thin — that is, the whole should be balanced.

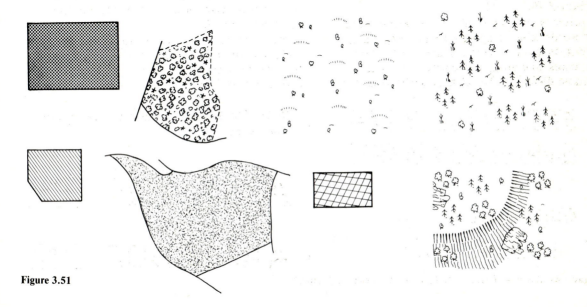

Figure 3.51

Part C — Height measurement

4 Bench marks

Introduction

Bench marks (BMs), which are the sources of vertical control, are fixed points whose heights relative to a datum surface have been determined by the use of a surveyor's level.

(Levelling equipment and procedures are described in §§ 5.1 to 5.7.)

Before continuing this section, it is suggested that the definitions of § 1.1 be revised (levelling, datum, level surface, etc.).

4.1 Ordnance datum and bench marks

Aim: *The student should be able to identify the datum to which Ordnance bench marks are referred.*

In § 1.1 a *datum surface* is defined as any arbitrary level surface to which the height of points may be referred, and a *level surface* is defined as one which, at all points, is normal to the direction of gravity.

Mean sea level (MSL) is such a datum surface, being the mean level of the sea at some selected place over a period of time. It is used by many national mapping organisations.

4.1.1 ORDNANCE DATUM

Ordnance datum (OD) is the mean level of the sea at Newlyn in Cornwall as calculated from hourly readings of the sea level recorded between 1915 and 1921 on a tide gauge situated in the Ordnance Survey tidal observatory on the end of the south pier at Newlyn. The datum surface is 4.751 m below a bench mark established at the observatory, known as the Ordnance Survey tidal observatory bench mark (*Figure 4.1*).

4.1.2 ORDNANCE BENCH MARKS (OBMs)

An *Ordnance bench mark* is a bench mark established by the Ordnance Survey, the height of the bench mark relative to Ordnance datum being known accurately. The Ordnance Survey have established OBMs all over the mainland and inshore islands of Great Britain, and levelling operations may be referred to these known heights.

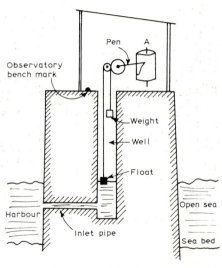

Figure 4.1 Diagrammatic section through a tidal observatory with recording at A on graph paper around a clock-driven rotating drum

4.1.3 DENSITY OF ORDNANCE BENCH MARKS

These are provided to meet normal user requirements, and their densities vary, according to the locality, from less than 300 m apart in city centres to over 1000 m in open rural areas.

4.1.4 TYPES OF ORDNANCE BENCH MARK

(a) *Cut bench marks:* These are the commonest OBMs, consisting of a horizontal line cut into a vertical surface such as a brick or stone wall. A government broad arrow is cut point upwards below the centre of the horizontal line (*Figure 4.2*).

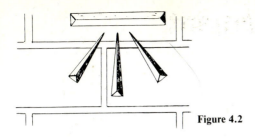

Figure 4.2

(b) *Fundamental bench mark (FBM):* These marks are placed on solid rock, at points approximately 40 km apart, and they act as control for the whole of the Ordnance Survey levelling network. Each mark consists of three reference points, two of these being in a buried chamber and not available for public use. The third mark is a brass or gunmetal bolt set on top of a low granite or concrete pillar, the height of

mushroom-headed bolts set horizontally in solid rock or concrete blocks (*Figure 4.4*).

(e) *Brass rivets:* These are occasionally used as an alternative to the cut bench mark where it is preferable to have the bench mark on a horizontal surface such as in a culvert or on the parapet of a bridge. The broad arrow is cut alongside, if convenient (*Figure 4.5*).

(f) *Pivot bench marks:* These are used on a horizontal surface, such as soft sandstone, where the insertion of a rivet would break away the stone. Instead of the rivet, a small hole or depression is cut to take a pivot which is a steel ball bearing of $\frac{5}{8}''$ (approx. 16 mm) diameter. In use, a pivot is placed in the depression and the staff held on top of the pivot. The published bench mark height refers to the top of the ball bearing (*Figure 4.6*).

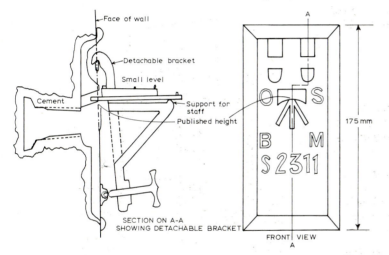

Figure 4.3

Figure 4.4 Plan view of bolt BM

the bolt being published. The bolt may be used by the site surveyor.

(c) *Bronze flush brackets:* Of lower order than the FBMs, these are levelling control points located on churches, large public buildings, etc., and cemented into the side of Ordnance Survey triangulation pillars. The published height is to the small horizontal platform at the point of the broad arrow marked on the face of the benchmark (*Figure 4.3*). A special staff support (also illustrated) is desirable when using this type of bench mark.

(d) *Brass or gunmetal bolts:* These are used by the Ordnance Survey when there is no suitable site for a flush bracket. They are 60-mm diameter

4.2 The publication of bench mark information

Aim: *The student should be able to use maps to obtain the position of (vertical) control points.*

Nearly all Ordnance Survey 1:1250 and 1:2500 scale maps, and large scale plans produced by

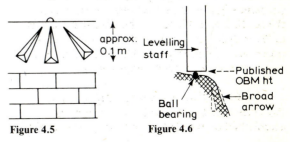

Figure 4.5   Figure 4.6

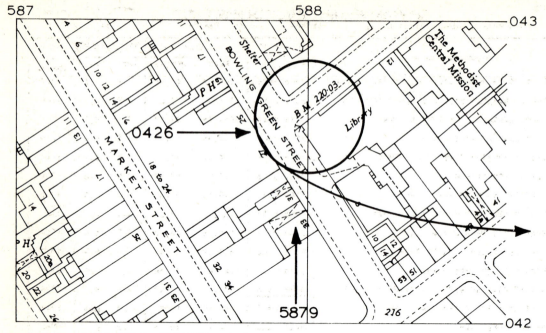

Figure 4.7 Part of OS 1:1250 plan SK 5804 SE, revised 1965

private survey companies and local authorities, indicate bench marks by means of a broad arrow, the head of which indicates the plan position (*Figure 4.7*). The height is usually given, and occasionally, the type of benchmark.

Ordnance Survey maps do not always contain the latest available levelling information. Further, it is not always possible to identify the date of the levelling, the type of bench mark, or the height of the mark above ground level. To overcome these problems, bench mark lists have been compiled (*Figure 4.8*), and these may be purchased direct from the Ordnance Survey by any site surveyor who needs complete and up-to-date levelling information.

A bench mark list is published for each square kilometre, and covering the same area which is (or would be) represented on a 1:2500 Ordnance Survey map. Both the map and the list are identified by the same national grid reference number. (*Site Surveying 3* give further information on the Ordnance Survey national grid referencing system.)

For each bench mark, the lists provide the following information:

(a) A brief description of the mark, together with its location,

(b) The full 10-m national grid reference,

(c) The height in metres and/or feet, with respect to Ordnance datum,

(d) The vertical distance in metres and feet of the bench mark above ground level,

(e) The date of levelling.

4.3 Temporary bench marks (TBMs)

Aim: *The student should be able to explain how temporary bench marks are established and give reasons for their use.*

A temporary bench mark is one which is set up by a site surveyor or engineer for use on a particular job. It may be set up by levelling from an Ordnance bench mark (*see* § 5.6). The advantage of establishing such a mark is that the levelling operations to be carried out on site may be referred back to the TBM without having to check back onto the Ordnance bench mark. TBMs, although semi-permanent, should be selected on some feature or point which is likely to remain stable throughout the construction programme.

ORDNANCE SURVEY BENCH MARK LIST

ALL BENCH MARKS ON THIS LIST FALL WITHIN KM SQ   SK5804   DATUM   NEWLYN   PAGE   2

DESCRIPTION OF BENCH MARK	NATIONAL GRID TEN METRE REFERENCE	ALTITUDE		HEIGHT OF BM ABOVE GROUND		DATE OF LEVELLING
		FEET	METRES	FEET	METRES	
POLYTECHNIC BLDG NO25 OXFORD ST NE FACE N ANG	5850 0404	208.81	63.65	1.8	0.5	1965
THE GLOBE PH NW SIDE SILVER ST SW FACE S ANG	5855 0451	213.28	65.01	2.2	0.7	1964
BLDG N SIDE NEWARKE ST S FACE SW ANG	5856 0414	213.64	65.12	2.1	0.6	1964
NO24 FRIAR LANE NW FACE N ANG	5858 0429	217.58	66.32	1.9	0.6	1965
BANK JUNC ST MARTINS GREYFRIAR N FACE NE ANG	5858 0441	214.39	65.35	2.2	0.7	1965
NO66 CHURCH GATE NE FACE	5863 0483	189.36	57.72	2.1	0.6	1964
NO12 MILLSTONE LANE NE FACE E ANG	5864 0428	218.38	66.56	1.9	0.6	1965
BLDG SW SIDE NEW BOND ST NE FACE E ANG	5865 0469	203.43	62.01	2.5	0.8	1964
NO1 NEWARKE ST S FACE W SIDE ENT	5868 0414	218.94	66.73	1.9	0.6	1964
CLOCK TWR EAST GATES E FACE NE ANG	5876 0402	200.17	61.01	1.6	0.5	1964
TH HORSEFAIR ST NW FACE N ANG	5878 0435	212.41	64.74	2.6	0.8	1965
LIBY NE SIDE BOWLING GREEN ST SW FACE W ANG	5879 0426	220.03	67.07	2.2	0.7	1964
NO15 WELLINGTON ST NE FACE 3M N ANG	5888 0408	217.12	66.18	1.6	0.5	1965
NO15 GRANBY ST NE FACE	5888 0440	206.48	62.94	1.2	0.4	1965
WALL SW SIDE NEW WALK NW SIDE ENT PH	5889 0402	217.56	66.31	1.0	0.3	1965

Date of issue   29/06/76   continued on page   3

The listed bench mark values may have been affected by local subsidence or physical disturbance since the date of levelling.

In England and Wales, BM's levelled before March 1956 are based on the second geodetic levelling, and those levelled afterwards on the third geodetic levelling. Differences between altitudes derived from these two levellings may be found due to the differing adjustments, but for most practical purposes these are insignificant. An estimated correction to relate one levelling to the other, by areas, is available from the Director General, Ordnance Survey.

In Scotland BM's are based on the second geodetic levelling, with additional heights derived from the third geodetic levelling in areas not previously levelled.

To convert metres to feet, divide the metric value by 0·3048.   © Crown Copyright: Unauthorised copying infringes Crown Copyright.

Figure 4.8

5 Levelling

Introduction

In § 1.1, *levelling* was defined as the art of determining the heights of different points relative to a datum surface by means of an instrument known as a *surveyor's level*.

Much of the site surveyor's work will be concerned with determining heights. This is not a difficult task, especially with the large number of bench marks which have been established by the Ordnance Survey. The basic equipment required includes a surveyor's level and a levelling staff, and over the years manufacturers have produced a very wide range of these items, many of them very similar to one another. An example of each of the different types of equipment with which the user should become familiar will be described, and provided that the user thoroughly understands how to use one particular level then he should have no difficulty in handling a similar instrument produced by another manufacturer.

5.1 The level

Aim: *The student should be able to describe the traditional levelling methods and instruments, including spirit and water levels, dumpy, tilting and automatic levels, and the site operative's automatic level (Contract etc.).*

A simple and effective device for defining a horizontal line is the *spirit level tube (Figure 5.1)* which was developed over 300 years ago. This consists of a slightly curved glass tube part-filled with a liquid before being sealed. The bubble of air remaining in the tube seeks the highest part of the tube. This 'bubble tube' may be seen in any builder's level (*Figure 5.2*).

If the device is attached to a straight edge it

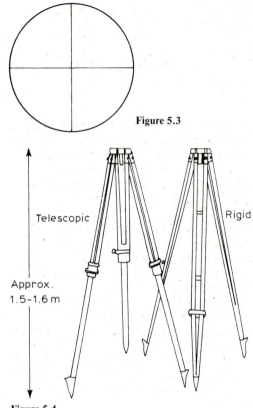

Figure 5.3

Figure 5.4

may be used to transfer heights over a distance of a metre or two. The transfer distance could be increased by attaching 'sights' to the straight edge but the range of the naked eye would still be very limiting. The obvious way to increase the sighting range is to attach the bubble tube to a telescope within which are crossed lines which act as a 'sight' for aiming (*Figure 5.3*). This line of sight through the telescope is known as the 'line of collimation' and it should be parallel to the bubble tube axis.

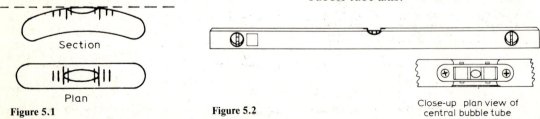

Figure 5.1

Figure 5.2

Close-up plan view of central bubble tube

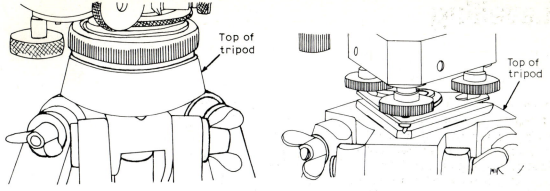

Figure 5.5

Figure 5.6

The telescope must be capable of being rotated in a horizontal or near-horizontal plane, and this is achieved by mounting it on a *levelling head* (also known as a *'tribrach'*) which is itself mounted on a tripod (*Figure 5.4*) to provide a stable platform at eye level. Two types of levelling head are commonly available, these being known as 'ball-and-socket' (*Figure 5.5*) or 'three-footscrew' (*Figure 5.6*), both of which enable the telescope to be tilted. Hence, the bubble can be 'centred' (*Figure 5.1*) and the line of collimation made horizontal.

This form of construction (a tripod-mounted telescope instrument) is the basis of the surveyor's level, of which there are three distinct types in current use, the *dumpy level,* the *tilting level* and the *automatic level.* These are described in the following sections, together with other instruments still commonly used for levelling work on site.

5.1.1 THE DUMPY LEVEL

The *dumpy level* came into use during the development of the railways, and the term 'dumpy' derives from the fact that its telescope was much shorter and thicker than those of the exisiting levels of the time. It should be noted that many building site operatives very often incorrectly refer to *any* small surveyor's level as a 'dumpy'.

The principal features of a typical instrument are shown in (*Figure 5.7*).

It is important to note that the telescope in this instrument is rigidly attached to a vertical axis spindle which rotates within and above the levelling head.

There are two essential requirements for accurate work:

(a) the bubble tube axis must be at right angles to the vertical axis, and

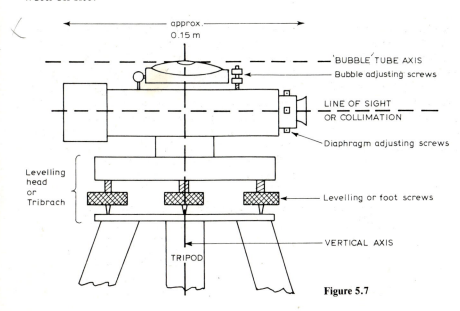

Figure 5.7

63

(b) the line of collimation must be parallel to the bubble tube axis.

The use of the bubble adjusting screws (seen in *Figure 5.7*) is explained in § 5.3.2(c).

As well as the features shown, dumpy levels have a means of focusing the 'cross-hairs' (effectively the internal sights) and the object being viewed. In addition, many levels have a clamp and slow-motion screw for locking and moving the telescope slowly when it is set up in a horizontal plane. Some levels also have a simple horizontal circle which may be useful on occasion when setting out, and a small circular bubble for the approximate levelling up of the instrument. The word *levelling* has already been defined, but the term 'levelling up' is applied to the operation of setting up the level in a horizontal plane.

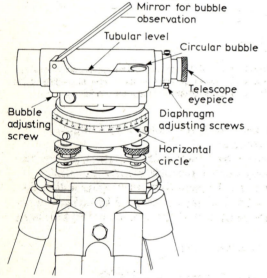

Figure 5.8 Wild NK 01 dumpy level

Figure 5.8 illustrates a typical dumpy in current use, the Wild NK 01 produced by Wild Heerbrugg Ltd. of Switzerland. (Note that the name 'Wild' is pronounced 'vilt', rhyming with tilt.)

The dumpy is obsolescent, few now being produced, but it can be a very useful instrument when many readings are to be taken from one instrument position.

5.1.2 THE TILTING LEVEL

The *tilting level*, first produced about 60 years ago, is a more advanced design than the dumpy, and it is generally quicker in use due to the shorter time required to set it up and prepare it for use.

The principal constructional features of the tilting level are shown in *Figure 5.9*. The telescope, bubble tube and diaphragm ('cross-hairs') are similar to those of the dumpy but the levelling head may be of either the ball-and-socket or three-footscrew types. (Note that very old instruments, dumpies, may be encountered with *four* footscrews, but all footscrew instruments now being made in Europe have three footscrews.)

Unlike the dumpy, the telescope of the tilting level is not rigidly fixed to the levelling head, but instead is supported on a pivot which allows the telescope to be tilted at an angle to the levelling head — hence the use of the name 'tilting'.

As with the dumpy, the bubble axis and the line of collimation should always be parallel. The instrument is levelled up approximately by the footscrews (or ball-and-socket if fitted) as judged by reference to the small circular bubble. Each time a reading is required, the instrument is levelled exactly by means of the fine-setting screw

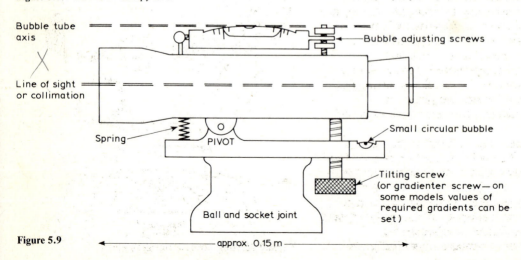

Figure 5.9

(the 'tilting screw') and the main bubble. This arrangement obviates the tedious levelling up needed with the dumpy level. *Figure 5.10* and *5.11* illustrate typical tilting levels in use today.

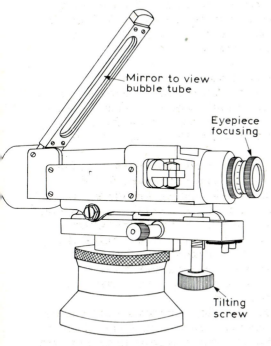

Figure 5.10 MOM Ni-E3 builder's level (with horizontal circle) produced by MOM Budapest

5.1.3 THE AUTOMATIC LEVEL

Automatic levels, developed and produced over the last 30 years, do not rely on a spirit level to attain a horizontal sight. Instead, these instruments typically make use of an arrangement of reflecting prisms fitted within the telescope barrel. These prisms are so arranged that they are in the line of sight from the surveyor's eye to the distant staff, and as the telescope is tilted the prism arrangement adjusts its position automatically to compensate for the deviation from the horizontal. The design and arrangement of the *compensating unit* (or *pendulum compensator*) differ between the various models produced.

The telescope is rigidly fixed to the vertical axis, as in the dumpy level, and the levelling head may be of either the three-footscrew or ball-and-socket types. A small circular bubble is fitted for approximately levelling up the instrument after its attachment to the tripod. When the instrument has been approximately levelled up the compensator unit operates automatically, its

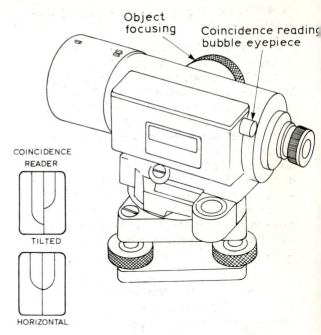

Figure 5.11 Watts SL432 engineer's level produced by Hall & Watts Ltd. The left-hand diagram shows how both ends of the bubble are seen side-by-side through the reading eyepiece

movement being slowed down by damping mechanisms.

Despite the damping device, vibrations due to blustery weather conditions and occasionally traffic and site plant may make sighting difficult. This vibration may be restricted by laying a hand lightly on the tripod, but remember that this must

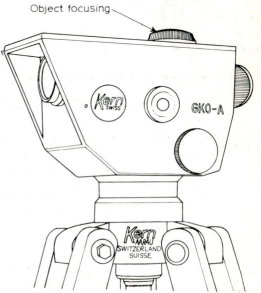

Figure 5.12 Kern GKO-A construction level

never be done with the other types of level, since it would disturb their bubble settings. Some early instruments had problems with excessive friction at the pivots of the suspended prisms, but these now appear to have been overcome.

Fieldwork may be carried out about twice as fast with the automatic level, as compared with the other types, and fewer mistakes occur because there is no bubble to be continually checked and adjusted. In addition, automatic levels give a right-way-up image while the majority of dumpy and tilting levels give an inverted image. Automatic levels are more expensive but, since their advantages far outweigh their disadvantages, the site surveyor may expect to see more and more of these levels in use.

Figure 5.12 illustrates one of the many automatic levels in use today.

5.1.4 OTHER LEVELS
The *site operative's automatic level* is not classed as a surveyor's level, but it is fast in use, convenient for small jobs and easily used by site operatives.

The construction incorporates a pendulum mirror system, defining a horizontal line of sight when the instrument is approximately levelled up on its lightweight tripod. Due to its lack of magnification, the sighting range is normally limited to approximately 30 m, and at this range manufacturers quote an accuracy of 6 mm. Slope attachments are available for setting out particular gradients, and some models may be mounted directly on top of brick courses for direct control of construction heights.

A *water level*, as described in § 1.2.2(d), is illustrated in *Figure 5.13,* this example being the Aqualev produced by Austin & Trimingham of London. Again, this is not a surveyor's level, but is widely used by site operatives in building work. It is a very precise method of transferring heights, typical applications in building work including checking the levels of brick courses, concrete shuttering, suspended ceilings, floor screeds, etc.

5.2 The staff and ancillary equipment
Aim: *The student should be able to read a metric levelling staff.*

5.2.1 THE LEVELLING STAFF
As explained earlier, the vertical distance between a point of known height (or a point whose height is to be determined) and the line of collimation is measured with a levelling staff. In site surveying, staves are usually made of wood but sometimes are of aluminium alloy. They are produced in a variety of lengths, possibly the 4-m is the more generally preferred length for site surveying. The construction of the staff may be telescopic, rigid one-piece, or of folded or jointed sections, the most popular in the United Kingdom being the telescopic type.

The graduation pattern used on the face of the staff may also vary, that illustrated in *Figure 5.14* being the 'E' pattern specified by BS 4484: Part 1: 1969. 'E-and-checkerboard' patterns are popular in Europe, the Wild version being considered by

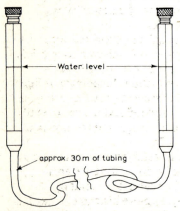

Figure 5.13

Figure 5.14

some surveyors to be easier to read than the British Standard pattern illustrated.

5.2.2 THE HAND OR STAFF BUBBLE
This circular spirit level (circular bubble) has been referred to in § 2.1 and shown in *Figure 2.18*. It is used to maintain the verticality of the staff during observations. It may either be attached to the back of the staff or held against a corner of the staff (*Figure 5.15*).

Figure 5.15

5.2.3 THE STAFF SUPPORT
The *staff support*, known also as a change plate, triangular plate, shoe, and (colloquially) as a crow's foot, is usually a triangular steel plate with a raised centre and the three corners turned down. A length of chain is attached for carrying and for removing the plate from the ground. The staff support is used on soft ground to provide a stable base for the staff and prevent it from sinking into the ground (*Figure 5.16*).

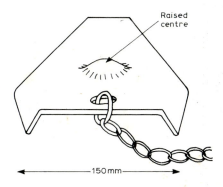

Raised centre

150mm

Figure 5.16

5.2.4 DETACHABLE BRACKET
This refers to the detachable bracket used with a bronze flush bracket bench mark (§ 4.1.4(c) and *Figure 4.3*).

5.3 Levelling fieldwork

Aim: *The student should be able to use levelling instruments and check their accuracy.*

5.3.1 PRELIMINARY TASKS
The following tasks should be carried out before levelling proper is commenced:

(a) Obtain the relevant bench mark information (§ 4.2).
(b) Inform the staffholder as to his duties (§ 5.3.4).
(c) Check the accuracy of the level to be used, if necessary (§ 5.3.5).
(d) Locate the bench marks to be used.
(e) Decide whether one or more instrument stations will be needed.
(f) Make preliminary decisions as to appropriate locations for the staff and instrument stations. In some cases, for example in establishing temporary bench marks, the instrument stations must be decided before starting (*see* § 5.6 concerning applications of levelling). The selection of the staff positions is often left to the staffholder to decide.

The location of staff and instrument stations will of course vary with the task to be carried out, the type of equipment in use, and the climatic and environmental conditions. However, the staff and instrument should always be set up on firm ground if this is at all possible, and the task should be carried out with as few instrument stations as possible, within the limitations on observing imposed by the length of the levelling staff and the horizontal distance. Distances between instrument and staff should be kept uniform, as far as possible, especially when levelling over long distances, or levelling up or down steep gradients, or when establishing temporary bench marks.

The purpose of equalising these distances is to minimise instrument errors, referred to in § 5.3.5, and also to reduce the effect of the Earth's curvature and the refraction ('bending') of light by the Earth's atmosphere. In site surveying, the effects of curvature and refraction are generally so small that they may be ignored for practical purposes (*Figure 5.17*).

The ideal length of sight between instrument and staff is from 45 to 60 m, longer sights tending to lead to inaccuracies in reading and shorter sights implying more instrument stations and a more costly job in terms of time and

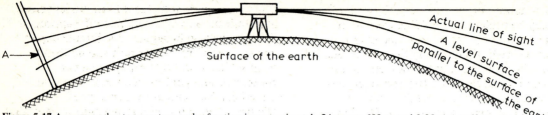

Line of sight parallel to the bubble axis

Actual line of sight

A level surface parallel to the surface of the earth

Surface of the earth

A —

Figure 5.17 A — error due to curvature and refraction is approximately 24 mm at 600 m and 0.25 mm at 60 m

money. The length of the levelling staff will impose restrictions on the length of sight on steep slopes, especially if equal sight lengths are to be maintained. Further, it is best to avoid reading the lower 0.2 m of the staff since refraction has its greatest effect near the ground. To maintain equal sights, it may be necessary to level up or down a hill in a zig-zag pattern in plan, or alternatively to select the high spots, if any, as instrument stations. Grazing rays should be avoided when passing over the crest of a hill (*Figure 5.18*).

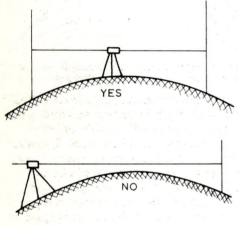

YES

NO

Figure 5.18

5.3.2 SETTING UP THE LEVEL (TEMPORARY OR STATION ADJUSTMENTS)

The operations involved in the *temporary adjustments* depend upon the type and make of the level and the tripod, the following sections describe typical outline procedures.

Note: These temporary adjustments must be carried out each time the equipment is used.

(a) Setting up the tripod

The modern tripod is of 'framed' construction, either 'rigid' or 'telescopic' (*Figure 5.19*). The rigid form is the more stable in use but the telescopic type is generally adequate and the

preferred type for site levelling operations. The telescopic type is easier to transport, more flexible in use, particularly on sloping ground or if the surveyor is either very short or very tall!

Tripods are generally made of thoroughly seasoned wood, with metal fittings, and when in continuous use under certain climatic conditions looseness may occur between the wooden and metal parts. All fittings should be checked from time to time and tightened as necessary, since loose fittings will cause difficulty in setting up the level and cause observational errors.

approx. one metre

approx. 1 m

Figure 5.19

The instrument position having been selected and the protective cap and fixing strap removed, the tripod legs should be extended so that when the level has been placed on the tripod the eyepiece of the telescope will be at a convenient viewing height. When adjusting the leg lengths, the clamps should be 'finger tight' only, and it must be remembered that spreading the legs and treading them in reduces the operating height of the tripod.

The tripod should now be placed in the chosen position, no attempt being made to set it up exactly over a ground mark since this is not necessary in levelling. On level ground each foot of the tripod should be equidistant from the others at about one metre apart (*Figure 5.19*). This may be achieved by holding the tripod vertical, with its top uppermost and over the position selected, then placing the foot of one leg in or on the ground about half a metre to the front and pulling the other two legs back and away on each side so that the feet form an equilateral triangle at ground level. On sloping ground one leg should point uphill with the other two downhill and at approximately equal ground heights. If the ground is soft, the feet of the tripod should be trod well into the ground using the lugs fitted to its legs and keeping one's inside leg against the tripod leg. On hard surfaces, to avoid the tripod slipping, its feet should be placed in small indentations in the ground, or in joints between paving slabs, if at all possible.

When the feet have been finally positioned the top of the tripod should be approximately horizontal, and it may be adjusted as necessary by extending or shortening one or two of the tripod legs. A final point to note is that the surveyor should attempt to set up the tripod so that he does not need to straddle one leg when observing, since this invites kicking the tripod which generally results in having to repeat the observations already made. Where tripods have wing nuts fitted at the tops of the legs, these should be clamped finger tight on completion of setting up.

(b) Attaching the instrument to the tripod

Before the instrument is removed from its case, a careful note should be made of exactly how it lies in the case, otherwise there may be problems when replacing it. The instrument should next be placed on top of the tripod and screwed finger tight in position. (With some models, the levelling head screws directly onto the top of the tripod, in other types a captive bolt in the head of the tripod is screwed into the base of the levelling head.)

Where a level is fitted with a ball-and-socket joint, the small circular bubble on the instrument should be floating centrally as the final turns of the tightening screw are made.

Finally, the telescope cap (if fitted) should be removed and returned to the case which should be closed and placed out of the way, an ideal location being under the tripod.

Note: When the level is being replaced in its case, the three footscrews should be returned approximately to their central positions.

(c) Levelling up the dumpy level

The telescope must be turned around its vertical axis until the axis of the bubble tube appears to lie parallel to an imaginary line joining *any* two of the footscrews. The bubble should then be 'centred in its run' by turning these two footscrews antagonistically, i.e. in opposite directions, but both at the same speed (*Figure 5.20*). It is useful to remember, here, that the 'bubble follows the left thumb'. The centring of the bubble is a delicate operation, and while waiting for the bubble to settle between adjustments neither the hands nor any other part of the body or clothing should be touching the instrument or the tripod.

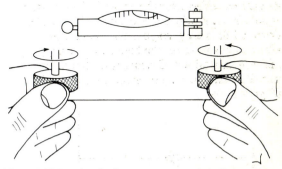

Figure 5.20 Turning the footscrews antagonistically to centre the bubble. Note: the bubble follows the left thumb

When the bubble is centred, the telescope should be rotated again about its vertical axis through an angle of approximately 90°, to right or left as convenient, and then the bubble carefully centred again using the third footscrew (the one *not* used in the previous centring). When the bubble is central, the instrument is approximately level. The telescope should then be returned to its *original* position and the bubble centred again by using the *original* two footscrews.

To check the accuracy of the levelling up, the telescope should be turned through 180° in plan and, when the bubble has settled its position in respect to the graduations on the bubble tube noted carefully. If the bubble has remained central, where it was before rotating through 180°, then the telescope should be turned through 90° and the bubble centred again using the third footscrew, if necessary. The instrument can now be considered to be properly levelled up, as may be tested by turning it through any angle and observing whether the bubble remains central. If it does not, then the *whole* of the levelling up procedure as described must be repeated.

If, as is mostly the case, it is found that the bubble moves off centre when the instrument is turned through 180°, then the bubble tube requires adjustment, as explained in § 5.3.5(a) However, if the bubble moves off centre but remains 'floating', it is still possible to level up the instrument.

To do this, the different positions of the bubble in its tube should be noted. In the example in *Figure 5.21*, the bubble is shown to

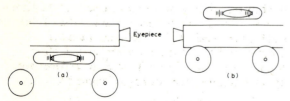

Figure 5.21 (a) Bubble central. (b) Bubble has moved two divisions towards the eyepiece

have moved through two divisions towards the eyepiece. The mean position, therefore, is *one* division towards the eyepiece, and the bubble should be set to this position using the original two footscrews. The telescope should then be turned through 90° and the third footscrew used to set the bubble in the mean position. The instrument should now be level. The levelling up should be tested by turning the telescope through any angle and noting whether the bubble remains stationary at, in this case, one division towards the eyepiece (*Figure 5.22*). If the bubble does not remain stationary then the whole of the levelling up procedure must be repeated from the beginning.

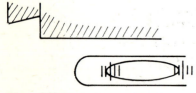

Figure 5.22

(d) Levelling up tilting and automatic levels
With these instruments, the levelling up is a much quicker and easier process, since all that is necessary is to get the small circular bubble approximately central (*Figure 5.23*).

If the levelling head is of the three-footscrew type, then the three footscrews should be used as necessary to centre the bubble, using a technique similar to that used for the dumpy — in this case, however, there is no need to turn the instrument in plan.

Where a ball-and-socket levelling head is fitted, the clamping ring or fastening screw (as

appropriate) should be eased off with one hand, while the other hand holds the telescope and tilts it as necessary until the circular bubble is central. Finally the ball-and-socket should be clamped without disturbing the centring of the bubble.

In the case of an automatic level, if the instrument is in adjustment the compensator unit should now ensure that the line of collimation is horizontal. With the tilting level, the tilting screw is used to centre the main bubble immediately before each observation.

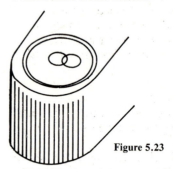

Figure 5.23

(e) Focusing and the elimination of parallax
As mentioned in § 5.1.1, both the cross-hairs and the object to be viewed must be focused, each being a separate operation. The cross-hairs must be focused first, and this should not need alteration during a day's work. The object (the levelling staff), however, must be re-focused each time the distance between the instrument and the staff is altered. Accurate focusing is essential, otherwise the object will appear to move up and down the cross-hairs if the surveyor moves his head. This movement between the object and cross-hairs is known as *parallax*, and it is something which no one can positively check since each surveyor's eyes are slightly different — only careful practice will ensure eventual elimination of parallax by the observer.

Focusing the cross-hairs: The telescope should be pointed at a light background, such as the page of a field book, and its eyepiece turned until the cross-hairs are brought into sharp focus. If in doubt as to which way to turn the eyepiece, it should be turned fully clockwise, then slowly 'unscrewed' anti-clockwise until the cross-hairs appear to be clear, sharp and black against the light background. As an alternative to using a light background, the object may simply be put out of focus by turning the focusing knob or screw until only the cross-hairs are seen in the field of view, then focusing the cross-hairs sharply with the eyepiece.

Focusing the object: Assuming that the cross hairs have been correctly focused with the eyepiece, the telescope should first be pointed on the distant staff using the open sights (if fitted), then the object brought into sharp focus using the focusing knob or screw. A check for parallax should be made by moving the eye slightly up or down, when the distant object and the cross-hairs should appear to be 'glued together'. Any relative movement between them indicates that the adjustment is incorrect and the check should be repeated as necessary. The focusing knob or screw is generally on the top or the right-hand side of the instrument as seen by the observer when looking at the eyepiece (*Figures 5.11 and 5.12*).

5.3.3 READING THE STAFF

The actions to be carried out in reading the staff vary slightly according to the type of level being used.

Using the *dumpy level*, check that the bubble is still central (or in its 'mean' position) then read the staff at the intersection of the cross hairs and book the reading (see § 5.5 for level booking).

Using the *tilting level,* centre the main bubble using the tilting screw, then read and book as above.

Using the *automatic level*, check that the compensator unit is working then book the staff reading as above. the compensator may be checked by a quick, light pressure on the tripod with one hand, and if the unit is operating correctly then the cross-hairs will appear to move slightly but will return to their original position rapidly.

As a general rule the staff should be read to 0.001 m, but in site surveying estimation to 0.002 m will often be adequate. Most modern levels do not have simple cross-hairs, but instead they are equipped with a more elaborate pattern

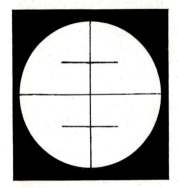

Figure 5.24

(*Figure 5.24*). The upper and lower 'hairs' are termed *stadia lines,* the whole arrangement forming the *graticule* or *reticule*. The stadia hairs may be used to determine the distance between the instrument and the staff, correct to within 0.2 m over the recommended maximum sighting distance of 60 m, an accuracy more than adequate for maintaining equal sights.

The instrument manufacturers have positioned the horizontal lines of the reticule so that they are equally spaced, and in such a manner that if the staff is read against all three hairs, then the horizontal distance from instrument to staff is equal to 100 times the staff intercept between the upper and lower hair readings (*Figure 5.25*). It

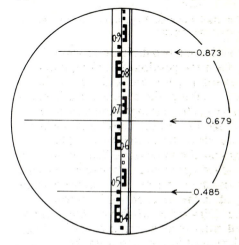

1 Has the centre hair been read correctly ?

$$
\begin{array}{r}
0.873 \\
+\ 0.485 \\
\hline
1.358 \div 2 = 0.679 \\
= \text{centre hair}
\end{array}
$$

2 What is the horizontal distance ?

$100 \times (0.873 - 0.485)$
$= 100 \times 0.388$
$= \underline{38.8\,m}$

Figure 5.25 Making use of the stadia hairs in levelling

should be noted that the sighting distances are not checked this way at every station, rather it is customary to rely on the staffholder to ensure equal distances by pacing.

The stadia hairs may also be used, particularly for an uncertain beginner, to provide a check on the centre hair reading, since the average of the upper and lower hair readings should be equal to the centre hair reading at the same point.

5.3.4 DUTIES OF THE STAFFHOLDER

A good staffholder is essential if the surveyor is to carry out the levelling accurately and quickly, and such an assistant will often anticipate the surveyor's requirements. Where possible, the site surveyor should advise the staffholder, before the commencement of the levelling, as to where staff stations will be required.

The staffholder should:
— ensure that the staff and ancillary equipment are in good working order;
— lay the staff on a clean, dry area of ground when it is not in use;
— carry the staff between stations, holding it vertically, and retracted if it is a telescopic type;
— check before setting up the staff that its zero end and the raised centre of the staff support are free from dirt;
— select a staff position which may be easily re-located later, if requested to do so;
— use a staff support, or a large stone, if the ground is not as firm as the surveyor would prefer for a staff position;
— extend or clamp the staff as necessary, ensuring that this has been done correctly;
— understand the simple system of hand signals used for communication in the field (as in chain survey);
— place the zero end of the staff gently onto the selected staff position, as required;
— hold the staff vertically for readings, keeping the hands off its face so as to avoid dirtying it or obstructing vision;
— check that the staff remains vertical during readings by occasionally glancing at the staff bubble;
— move the staff or its support only when instructed to do so;
— inform the surveyor if the staff (or its support) is displaced or tending to sink (possibly under its own weight or through the staffholder inadvertently leaning his weight on it;
— maintain equal distances (within about a metre) between an instrument and its foresight and backsight staff stations by careful pacing;
— wipe clean, and dry off the staff and ancillary equipment upon completion of the day's levelling.

5.3.5 PERMANENT ADJUSTMENTS OF THE LEVEL

Instruments should be checked for accuracy:
— when first received,
— whenever they have been severely jolted,
— at least once every six months, if in continual use, and
— at any time when the site surveyor thinks they may be in error.

If a level is found to be in error by an excessive amount (say, an error in reading of more than 2 mm per 30 m of sighting distance) then it should be adjusted. This type of adjustment is termed a 'permanent adjustment', and its details vary with the type of instrument in use.

(a) Permanent adjustment of the dumpy level
The dumpy level has two permanent adjustments:
(i) to make the bubble tube axis perpendicular to the vertical axis, and

(ii) the 'collimating adjustment' whereby the line of collimation is brought parallel to the bubble tube axis and hence at right angles to the vertical axis.

To carry out adjustment (i): The instrument should first be levelled carefully, as described in § 5.3.2, to obtain the mean position of the bubble. This has been done in the example in *Figure 5.22* where the position of the bubble appears to be correct and the graduations on the bubble tube appear to be incorrect. The situation may be corrected by using the bubble tube adjusting screws to reset the tube so as to centre the bubble exactly. With adjustment complete, the bubble should be re-checked and the adjustment repeated if necessary, *Figures 5.7* and *5.8* show typical examples of where the bubble adjusting screws may be located.

To carry out adjustment (ii): In applying this adjustment, it is common practice to test the state

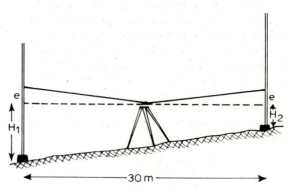

True height diff $= H_1 - H_2$
Error 'e' will be the same at both staff positions, if the instrument is set up half-way between the staves.
∴ diff in readings
$= (H_1 + e) - (H_2 + e)$
$= H_1 + e - H_2 - e$
$= H_1 - H_2$
$=$ true height diff

Figure 5.26

of the instrument by using two pegs, hence the method is known as the 'two peg test'.

On reasonably level ground two stable marks, typically wooden pegs, should be set up about 30 m apart with the instrument set up and levelled midway between them (*Figure 5.26*).

If a staff is held on each mark in turn, and the readings noted, then the difference between the readings will be the difference in height between the two marks, regardless of any error in the line of collimation.

The level should then be moved and set up again, still between the marks but positioned so that the *eyepiece* is only 25 to 50 mm away from the staff held on one of the marks. (*Figure 5.27*).

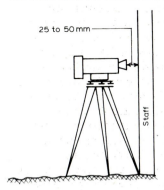

Figure 5.27

Having carefully levelled up the instrument, the staff should be read through the *objective* end of the telescope. In reading the staff this way, the surveyor may not be able to see the cross-hairs, but this is not important since the field of view will be very small and it is easy to judge the centre of the visible circle. A pencil point laid against the staff face (*Figure 5.28*) will aid in determining the reading, since the pencil will be visible through the telescope even if no figures can be seen. The pencil is moved until its point is central then the graduation at the pencil point is observed directly by eye.

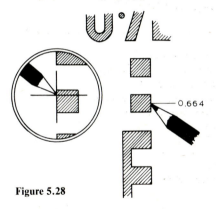

Figure 5.28

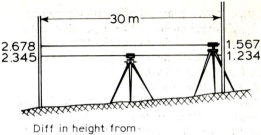

Diff in height from
1st pair of readings
= 2.345 - 1.234 = 1.111 m
Diff in height from
2nd pair of readings
= 2.678 - 1.567 = 1.111 m
∴ instrument is correct
Calculations can be checked by:
left - hand readings
= 2.678 - 2.345 = 0.333 m
right - hand readings
= 1.567 - 1.234 = 0.333 m

Figure 5.29

Finally, the staff should be moved to the other mark, the instrument focused, and the staff read again. If the instrument is in adjustment the difference between these two staff readings should be the same as the difference between the first two staff readings, any discrepancy indicating the existence of collimation error. (*Figures 5.29, 5.30*).

It should be noted that there are various ways of carrying out the two peg test, as will be found in different textbooks and manufacturers'

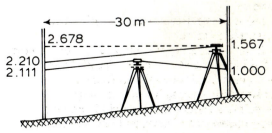

Diff in height from
1st pair of readings
= 2.111 - 1.000 = 1.111 m

Diff in height from
2nd pair of readings
= 2.210 - 1.567 = 0.643 m
∴ instrument is in error

True diff in height (from
1st pair of readings)
is 1.111 m
∴ if 1.567 is correct, then
the final rdg must be
1.567 + 1.111 = 2.678 m
Check 2.678 - 2.111 = 0.567 m
and 1.567 - 1.000 = 0.567 m

Figure 5.30

handbooks. The method described here is simple and it overcomes the problem of the long minimum focusing distance of some dumpy levels while simplifying the calculations which may be needed. If any adjustment is required it will involve only the most minute movement of the diaphragm, and the reading taken with the instrument close to the staff will not be changed in any measurable amount.

From *Figure 5.30* (ignoring *Figure 5.29*) it can be seen that the final reading in the example should be 2.678. If the bubble adjustment has already been carried out, then the error must lie in the vertical positioning of the cross hairs. The cross hairs may be raised or lowered as needed to eliminate the error by alternately loosening and tightening the diaphragm adjusting screws (*Figure 5.7, 5.8*) to make the staff reading equal to the calculated value of 2.678 m.

When adjustment is complete, the test should be repeated as a check and further adjustment made if necessary.

(b) Permanent adjustment of the tilting level
The tilting level has only one permanent adjustment — to make the bubble tube axis parallel to the line of collimation. The adjustment is tested in exactly the same way as for the collimation adjustment of the dumpy level, described above and illustrated in *Figure 5.29, 5.30*.

To carry out the adjustment, the correct staff reading must be calculated (2.678 in example *Figure 5.30*) then the tilting screw used to bring the cross-hairs onto this calculated value (*Figure 5.9, 5.10*). This operation will, in turn, cause the bubble to move off centre, and it should then be centred again using the *bubble tube adjusting screws*.

The test and adjustment should be repeated as necessary.

It is important to note the different procedures: In dumpy level collimation adjustment, the line of collimation is moved with respect to the telescope, while in the tilting level the position of the bubble tube is adjusted.

(c) Permanent adjustment of the automatic level
The automatic level also has only one permanent adjustment — to make the line of collimation horizontal when the small circular bubble is centred.

The adjustment is tested in exactly the same way as for the collimation adjustment of dumpy or tilting levels, as described above.

To carry out the adjustment, the correct staff reading must be calculated as before (again, 2.678 in *Figure 5.30*) then the cross-hairs brought exactly onto this calculated value on the staff. The method of moving the line of collimation to get the cross-hairs onto the reading, however, varies between different instruments, and it is necessary to refer to the maker's handbook for the model in use. Some instruments are adjusted by moving the diaphragm in the same way as with the dumpy, while in others it is the compensator unit which is adjusted. In still others, both items must be adjusted.

It should be noted that in some automatic levels the eyepiece and the objective lens do not lie in the same horizontal plane (*Figure 5.31*),

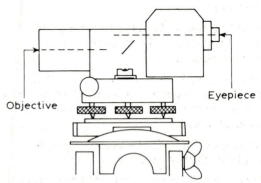

Objective   Eyepiece

Figure 5.31

hence the method of looking at the staff through the objective lens is not acceptable. In the instrument illustrated, the Sokkisha B2 A, the shortest focusing distance is 1.8 m, thus the level must be set up in the normal manner but at least 1.8 m from the near staff. If any adjustment is needed it will mean that the near reading may have been altered also when the cross-hairs are moved, but any such movement should be minimal.

As in all cases, the adjustment should be checked and repeated as necessary.

5.4 Terms used in levelling
Aim: *The student should be able to define the terms used in levelling.*

The following terms, commonly used in site levelling work, are illustrated in the line of levels (*series levelling*) shown in *Figure 5.32*.

Backsight: The first reading taken from any instrument position.

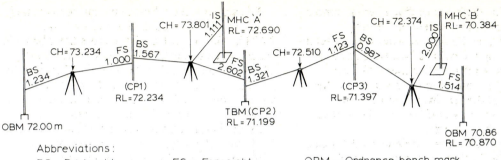

Figure 5.32

Abbreviations:

BS - Backsight   FS - Foresight   OBM - Ordnance bench mark
CH - Collimation height   IS - Intermediate sight   RL - Reduced level
CP - Change point   MHC- Manhole cover   TBM - Temporary bench mark

Foresight: The last reading taken from any instrument position.

Intermediate sights: Readings which are neither the first nor the last to be taken from an instrument position.

Change point (CP): A staff position at which first a foresight from one instrument position and then a backsight from another instrument position are taken.

Reduced level (RL): The calculated height of a point above or below a datum surface.

Collimation height: The calculated height of the line of collimation above or below the datum surface.

Rise and fall: The vertical distance between two consecutive staff positions is either a 'rise' or a 'fall', a *rise* being a positive difference (the second point higher than the first) and a *fall* being a negative difference (the second point lower than the first).

5.5 Level booking

Aim: *The student should be able to record and calculate reduced levels by 'rise-and-fall' and 'collimation-height' methods, and apply the required checks.*

It may be possible to take all the level readings required on a site from a single instrument position (one set-up), or it may be necessary to set up the level at several different positions (*Figure 5.32*). Whichever occurs, the readings are recorded in a ruled *level book*, the actual form of ruling depending on the surveyor's preference. There are two patterns in general use, the 'rise-and-fall' and the 'collimation' type. (Note that some of the information concerning chain survey

booking, § 3.2, is also applicable to level booking.)

As in chaining, the site surveyor generally does the booking and also reduces the readings. It is best to reduce the levels as the levelling proceeds and check and complete the reductions before leaving the site.

5.5.1 THE RISE-AND-FALL METHOD

The line of levels shown diagrammatically in *Figure 5.32* is used in *Figure 5.33* to illustrate this method of booking.

Points to note in *Figure 5.33*:

Each line represents a staff position, and that position is identified in the 'remarks' column.

The first reduced level is entered from the given data, the remainder being calculated as follows:

Obtain the rises or falls, remembering that a rise or fall is the difference between the readings to two consecutive staff position, thus:

$$1.234 - 1.000 = +0.234 = \text{rise}$$
$$1.567 - 1.111 = +0.456 = \text{rise}$$
$$1.111 - 2.602 = -1.491 = \text{fall}$$
$$1.321 - 1.123 = +0.198 = \text{rise}$$
$$0.987 - 2.000 = -1.013 = \text{fall}$$
$$2.000 - 1.514 = +0.486 = \text{rise}$$

(**Note** carefully where these calculated values have been entered on the booking sheet, and note also that while each backsight or foresight is used only once, each intermediate sight value is used twice.)

Using the calculated rises and falls, calculate the reduced levels in succession from:

BACK SIGHT	INTER- MEDIATE	FORE SIGHT	RISE	FALL	REDUCED LEVEL	REMARKS
1.234					72.000	OBM St. Johns Church
1.567		1.000	0.234		72.234	CP1
	1.111		0.456		72.690	MHC 'A'
1.321		2.602		1.491	71.199	TBM (CP2)
0.987		1.123	0.198		71.397	CP3
	2.000			1.013	70.384	MHC 'B'
		1.514	0.486		70.870	OBM (70.86) The Ring Of Bells 'PH'
5.109		6.239	1.374	2.504	72.000	
6.239			2.504		−1.130 ✓	
−1.130 ✓			−1.130 ✓			

Figure 5.33

Reduced level = reduced level on previous line
+ rise between the two staff positions, or
− fall between the two staff positions

e.g. $72.000 + 0.234 = 72.234$

$72.234 + 0.456$
$= 72.690$
$72.690 − 1.491$
$= 71.199$
$71.199 + 0.198 = 71.397$ etc.

The calculations may be summarised as:

$(BS_1$ or $IS_1) − (FS_2$ or $IS_2)$ is a rise if positive, or a fall if negative
$RL_1 + rise_2$ or $− fall_2 = RL_2$.

The commonest mistakes which occur in level booking and reduction are arithmetical, hence every arithmetical operation must be checked. If the calculations above are correct, then:

The sum of the backsights − the sum of the foresights
= The sum of the rises − the sum of the falls
= The last reduced level − the first reduced level.

Note: *In practice* the correct routine is to calculate the rises and falls, then compare the difference between the backsight and foresight sums with the difference between the sum of the rises and the sum of the falls, to ensure that the rises and falls have been accurately computed. Only when this arithmetic has been checked should the reduced levels be calculated. Finally, the checking should be completed by comparing these differences with the difference between the first and last reduced levels.

The rise-and-fall method provides a complete check on the arithmetic of the reductions, but it must be appreciated that it does *not* check the accuracy of the actual observations. These may

BACK	INTER- MEDIATE	FORE	COLLIMATION	REDUCED LEVEL	REMARKS
1.234			73.234	72.000	OBM St. Johns Church
1.567		1.000	73.801	72.234	CP1
	1.111			72.690	MHC 'A'
1.321		2.602	72.520	71.199	TBM (CP2)
0.987		1.123	72.384	71.397	CP3
	2.000			70.384	MHC 'B'
		1.514		70.870	OBM (70.86) The Ring of Bells
5.109		6.239		72.000	
6.239				−1.130	
−1.130 ✓					

Figure 5.34

be checked or 'proved' only by levelling back to the opening bench mark or completing the line of levels onto another point of known height and comparing the calculated and the known heights.

It will be noted that in the above example there is a misclosure of 10 mm, since the calculated reduced level at the end of the line of levels is 70.870 while the given height is 70.86 m. The method of dealing with this is described in § 5.5.4.

5.5.2 THE COLLIMATION (HEIGHT) OR HEIGHT-OF-INSTRUMENT METHOD

The level bookings here are exactly the same as for the rise-and-fall method, the difference lying in the method of reduction. Again, the line of levels in *Figure 5.32* is used in *Figure 5.34* to illustrate booking and reduction by this method.

Points to note in *Figure 5.34:*

Again, each line represents a staff position, and that position is identified in the 'remarks' column. The first reduced level is again entered from the given data, but the remainder are calculated as follows:

Reduced level + backsight = collimation height, (collimation height being the height of the instrument line of collimation above datum.)
and
Collimation height − foresight (or intermediate sight, as relevant) = reduced level.

Thus:
72.000 + 1.234 = 73.234 (inst ht)
73.234 − 1.000 = 72.234 (RL of CP 1)

72.234 + 1.567 = 73.801 (inst ht)
73.801 − 1.111 = 72.690 (MHC 'A')
73.801 − 2.602 = 71.199 (RL of TBM)

71.199 + 1.321 = 72.520 (inst ht)
72.520 − 1.123 = 71.397 (RL of CP 3)

71.397 + 0.987 = 72.384 (inst ht)
72.384 − 2.000 = 70.384 (MHC 'B')
72.384 − 1.514 = 70.870 (RL of OBM)

(Again, the position where these values have been entered on the booking sheet should be noted carefully.)
The calculations could be summarised as:

RL + BS = CH,
$CH_1 - FS_2 = RL_2$, or
$CH_1 - IS_3 = RL_3$

Mathematical checks are that the sum of the backsights less the sum of the foresights should equal the last reduced level less the first reduced level, and this is often the only check applied.

Unfortunately, this does *not* check the calculation of the reduced levels of the intermediate sights. The full arithmetic check is so laborious that few individuals use it on site works, and it is not described here. A full description of the complete check may be found in *Basic Metric Surveying,* Whyte, W. S., 2nd edition, Butterworths 1976.

Again, it must be remembered that arithmetic checks do not check the accuracy of the actual field observations.

The collimation-height method is widely used, and often preferred, when levelling on site as it is considered to be quicker in use. The commonest errors in levelling, however, are arithmetical, particularly with individuals who seldom use a level. Accordingly, such persons, and beginners, would be best advised to use the rise-and-fall method to eliminate these mistakes.

5.5.3 CHECKING LEVELS EXTENDING OVER MORE THAN ONE PAGE

When level book entries extend over more than one page, each page should be checked separately when reducing, although with a greater use being made of pocket calculators this is perhaps not so important today. If each page is checked, and this is strongly recommended, then three or four lines must be left clear at the bottom of each page or booking sheet to allow for totals and differences. If a page is to be checked separately, then the entries on the page must commence with a *backsight* and end with a *foresight*. It will be evident that the last reading on a page will generally be *either* a reading to a change point *or* a reading to an intermediate sight.

If the last entry is for a change point, then the foresight reading to that point will be the last reading on the page, and the backsight reading to the same point must be entered as the first reading on the next page.

If the page ends at an intermediate sight, then it must be booked as a *foresight*, and repeated again as a *backsight* as the first reading on the next page.

Examples of these are shown in *Figure 5.35* where the centre booking sheet is a typical specimen of a page of rise-and-fall booking, with all the entries on one page.

The same readings are shown on the left, booked on two pages with the new page

Back	Inter	Fore	Rise	Fall	RL
4.50					94.20
	4.00		0.50		94.70
	2.05		1.95		96.65
3.32		0.42	1.63		98.28
	2.28		1.04		99.32
		1.54	0.74		100.06
7.82		1.96	5.86	0.00	94.20
1.96			0.00		5.86
5.86			5.86		✓

Back	Inter	Fore	Rise	Fall	RL
4.50					94.20
	4.00		0.50		94.70
	2.05		1.95		96.65
3.32		0.42	1.63		98.28
	2.28		1.04		99.32
0.26		1.54	0.74		100.06
	2.22			1.96	98.10
0.20		4.24		2.02	96.08
	1.80			1.60	94.48
	3.90			2.10	92.38
	4.10			0.20	92.18
8.28		10.30	5.86	7.88	94.20
10.30			7.88		-2.02
-2.02			-2.02		✓

Back	Inter	Fore	Rise	Fall	RL
4.50					94.20
	4.00		0.50		94.70
	2.05		1.95		96.65
3.32		0.42	1.63		98.28
	2.28	1.04			99.32
7.82		2.70	5.12	0.00	94.20
2.70			0.00		5.12
5.12	✓		5.12	✓	

Back	Inter	Fore	Rise	Fall	RL
0.26					100.06
	2.22			1.96	98.10
0.20		4.24		2.02	96.08
	1.80			1.60	94.48
	3.90			2.10	92.38
		4.10		0.20	92.18
0.46		8.34	0.00	7.88	100.06
8.34			7.88		-7.88
-7.88			-7.88		✓

Back	Inter	Fore	Rise	Fall	RL
2.28					99.32
0.26		1.54	0.74		100.06
	2.22			1.96	98.10
0.20		4.24		2.02	96.08
	1.80			1.60	94.48
	3.90			2.10	92.38
		4.10		0.20	92.18
2.74		9.88	0.74	7.88	99.32
9.88			7.88		-7.14
-7.14	✓		-7.14	✓	✓

Figure 5.35

commencing at a change point, while on the right the same readings are again spread over two pages but with the new page starting at an intermediate sight.

5.5.4 PERMISSIBLE ERRORS IN LEVELLING
The example level reductions in *Figures 5.32, 5.33* and *5.34* show a misclosure of 10 mm, or 0.010 m. The reductions were checked for arithmetical errors and found to be correct, hence the 0.010 m must be an error in the levelling and not in the calculations. Since there are always errors in survey, a limit has to be set for the *permissible* (acceptable) *error* in any levelling job. The actual error permissible depends upon the type of job and possibly the class of level being used. Levels have already been placed in one of three types — dumpy, tilting, automatic — but they are also classed by makers as being builder's, engineer's or precise, depending upon the quality of the optics and the precision of their construction. (Hence there are builder's dumpies, builder's tilting levels, and builder's automatic levels, etc.)

In site surveying, closing errors not exceeding $5\sqrt{s}$ mm are generally acceptable, where s = the number of instrument set-ups used. On longer lines of levelling a tolerance of $20\sqrt{K}$ mm where K = the length in kilometres of a single line of levelling is often used. In establishing temporary bench marks, however, $10\sqrt{K}$ mm would be preferred.

Commonsense must be applied in deciding whether a misclosure is acceptable or otherwise. As an example, if reduced levels were required for contouring only, then a misclosure of $100\sqrt{K}$ mm might be adequate, although such a standard should be discouraged in levelling generally (see *Site Surveying 3* for contouring). When the permissible error for a task has been exceeded, and the errors cannot be located, then the levelling must be repeated (*check levelling*). If at all possible it is best to carry out check levelling using the same change points as were used in the original levelling, since then the error may possibly be located with only one or two set-ups of the instrument.

Where a misclosure is within the limits of the permissible error for the task, then unless the error is extremely small it should be distributed uniformly through the levelling, applying corrections at each change point. *Figure 5.36* to

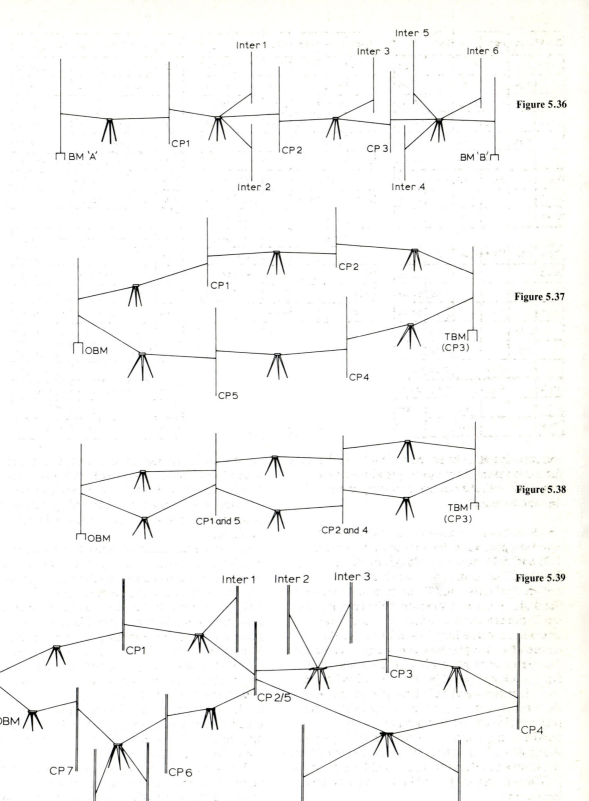

Figure 5.36

Figure 5.37

Figure 5.38

Figure 5.39

5.39 illustrate four typical lines of levelling, and the method of adjusting each is described.

Figure 5.36 shows a line of levels from bench mark 'A' to bench mark 'B'. Assuming a misclosure of 10 mm, then approximately $2\frac{1}{2}$ mm correction would be required at CP1, 5 mm at CP2, $7\frac{1}{2}$ mm at CP3 and 10 mm at BM 'B' (three change points and the end point, apply 10/4 mm at each CP successively). Since it is usual practice to work to the nearest millimetre, the practical corrections would be 2 mm at CP1, 5 mm at CP2, 7 mm at CP3 and 10 mm at BM 'B'.

Reduced levels at intermediates should be adjusted by the amount of correction applied at the preceding change point.

Figure 5.37 shows a closed levelling loop, run to establish a TBM at a point which forms CP3 in the loop. Assuming a misclosure of 8 mm in the loop, then a uniform distribution would require corrections of 8/6 mm at CP1, $2 \times 8/6$ mm at CP2, $3 \times 8/6$ mm at CP3, $4 \times 8/6$ mm at CP4, etc., since there are five change points and the end point making six points in all. A practical set of corrections would be: 1 mm at CP1, 3 mm at CP2, 4 mm at CP3, 5 mm at CP4, 7 mm at CP5 and 8 mm at the OBM.

Figure 5.38 shows a better way of establishing the TBM at point CP3, which is to make CP1 and CP5 coincident and similarly make CP2 and CP4 coincident. In this case, adjustment of the misclosure may be tackled in two different ways, although both should produce the same results.

One method is to regard it as a single loop, as in *Figure 5.37* and obtain two distinct heights for point CP1/5 (a height as CP1 and a separate height as CP5 'coming back') and two distinct heights for CP2/4 (a height as CP2 'going out' and a height as CP4 'coming back'), adjusting the whole circuit of six points. The adjusted levels of CP1 and CP5 may be meaned (averaged) to give an adjusted level for CP1/5 and the levels of CP2 and CP4 meaned to give an adjusted level for CP2/4.

An alternative method is to take the mean of the height differences between the OBM and CP1/5 and apply this to the known height of the OBM, thus giving the required reduced level for the point CP1/5. Similarly the mean differences of height between CP1/5 and CP2/4 may be applied to the reduced level of CP1/5 to give the required reduced level of CP2/4, and the same process continued to give the reduced level of the TBM. The advantage of this method is that the site surveyor will readily spot gross errors when visually inspecting what should be compatible readings.

Figure 5.39 shows a closed loop with intermediate sights, linked at its approximate centre point. This again may be tackled in two different ways, but the best approach is to treat it as two separate loops. The reduced levels would be calculated from the OBM through CP1, CP2/5, CP6, CP7 and back to the OBM., and the loop adjusted if found to be satisfactory. The remaining levels would be calculated and adjusted from CP2/5 through CP3, CP4 and back to CP2/5.

It would be a useful exercise at this stage to return to *Figure 5.32* and example bookings in *Figures 5.33* and *5.34* and adjust the reduced levels.

5.6 Applications of levelling
Aim: *The student should be able to undertake the different forms of levelling — flying, grid, cross-sectional and inverse.*

5.6.1 FLYING LEVELS — ESTABLISHING A TBM
Flying levels is the name given to a line of levels, without intermediate sights, running between two different bench marks or running in a circuit from one bench mark and back to the same bench mark. (*Figures 5.37, 5.38*) The procedure for establishing a temporary bench mark, using flying levels, is as follows:

(a) Obtain bench mark information (§ 4.2);

(b) locate the bench marks;

(c) select the temporary bench mark (§ 4.3);

(d) decide on the route the line of levels is to follow (§ 5.3.1);

(e) instruct the staffholder (§ 5.3.4);

(f) set up the level (§ 5.3.2);

(g) sight on to the opening bench mark, read and book the staff value (§§ 5.3.3, 5.5);

(h) signal the staffholder to move (the surveyor must not leave the instrument unattended, particularly where there is much vehicular traffic);

(i) sight the staff, take and book the foresight reading;

(j) move the instrument on to the next position (the instrument may be carried on its tripod if the legs are closed and it is firmly held in a vertical position, but if it is to be carried in a vehicle then it should be replaced in its carrying case and held on the lap of one of the passengers);

(k) sight the staff, take and book the backsight reading at the changepoint; and

(l) repeat the process until the temporary bench mark is reached, then close the line of levels onto another bench mark, or alternatively onto the original bench mark.

5.6.2. GRID LEVELLING

Grid levelling is the method used to obtain the heights of a rectangular network of ground points over a site (a 'grid' of levels over the area).

Grid levels may be used to determine the average height of a site, for contour interpolation when a contoured plan is required, and in the calculation of the volumes of earthworks. These later applications are covered in *Site Surveying 3*.

The procedure for carrying out grid levelling (*Figure 5.40*) may be summarised as follows:

(a) Obtain bench mark information for the area;

(b) Obtain a plan of the site (if there is no plan, a new survey may be required, but in some cases a plan is not needed);

(c) locate the bench marks to be used;

(d) decide the grid interval to be used and how the grid will be tied to the plan (the ground);

(e) mark out the grid on the ground;

(f) establish any necessary TBM(s); and

(g) level the grid points.

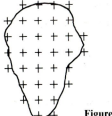

Figure 5.40

(a) Fixing the grid interval and tying the grid to the plan

Typical grid point interval may be 10 or 20 m, but any value between 5 and 50 m may be used, depending upon the nature of the site and the job. The larger the spacing the fewer points to locate and level, but generally the spacing is influenced most by the undulations of the ground and the accuracy required. When grid points are located on the ground, there should be no significant change of slope between any two adjacent grid points, so that the surface of any grid square should, for all practical purposes, approximate to a plane (*Figure 5.41*).

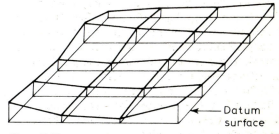

Figure 5.41

The grid should be tied to a straight line on the plan, often one of the chain lines is used if the site is being surveyed at the same time, but a straight portion of detail is suitable. The grid is often laid out using a tape and an optical square, particularly on smaller sites, and in this case if the offsets are to be kept short it is preferable that the grid is based on a line through the centre of the site (*Figure 5.42*). Offsets will often be

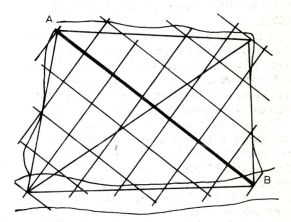

Figure 5.42

long by the standards of chain survey, and errors larger than those acceptable in chaining may occur, but these will not significantly affect the end product.

(b) Marking out the grid

The grid should be marked on the site with pegs, ranging rods or other marks. As an example, in *Figure 5.42*, where the grid is to be based on the line AB, then pegs would be placed at A and (for 20-m spacing) at 20-m intervals from A to B. At each of the pegs right angles would be raised with the optical square and further pegs placed at

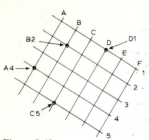

Figure 5.43

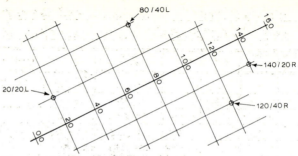

Figure 5.44

20-m intervals along these raised offsets. On a large job this may entail an enormous number of pegs, but this problem may be overcome by, instead, placing a double line of ranging rods along each of two adjacent sides of the grid, the staffholder then lining himself in by ranging with any pair of rods. With a large team of assistants the levelling and the setting out of the grid may be carried out at the same time, in which case pegs may be dispensed with. It is usual to extend the grid slightly beyond the site boundary.

Identifying the grid points
Grid levelling is likely to involve a very large number of intermediate sights, and a systematic approach to the identification of the points is essential in order to avoid any misunderstanding between the surveyor and the staffholder. Two recognised methods are in general use, the first being to allot identifying *letters* to the lines in one direction while allotting *numbers* to the lines in the other direction, then any point may be specified by the lines which intersect at that point, e.g. A4, B2, C5, D1 etc., (*Figure 5.43*).

The other common method, where the grid is based on a centre line, is to identify any point by its distance left or right of a point at a specified distance along the centre line (*Figure 5.44*). An advantage of this method is that any point on the ground can be identified, not just grid points, as for example the point 120/72.5R, meaning the point 72.5 m to the right of a point 120 m along the centre line.

A systematic approach is also necessary when considering the route to be taken by the staffholder as he works over the site. In *Figure 5.43* the staffholder could start at F1 then go to F2, F3, F4, F5, E5, E4, E3, E2, E1, D1, D2, and so on. A sketch of a site, together with the field book and reductions, is shown in *Figures 5.45* and *5.46*. The route taken by the staffholder will be clear from the bookings.

(d) Determining the average height of the site

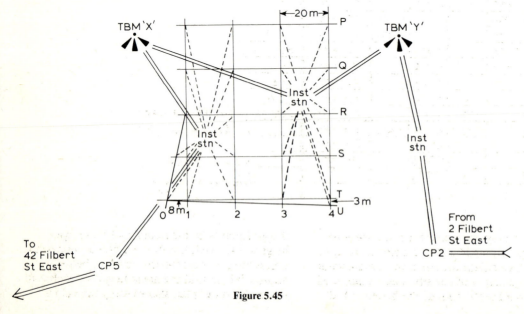

Figure 5.45

Date 15th May 1980   Levels taken for LRI Car park site

From OBM 2 Filbert St. East   To OBM 42 Filbert St. East

BACK SIGHT	INTER-MEDIATE	FORE SIGHT	RISE	FALL	REDUCED LEVEL	REMARKS
1.260					59.230	OBM 2 Filbert Street East
0.010		1.110	0.150		59.380	CP 1
0.313		0.145		0.135	59.245	CP 2
1.502		1.825		1.512	57.733	TBM 'Y' (CP3)
	1.960			0.458	57.275	P4
	1.885		0.075		57.350	Q4
	1.959			0.074	57.276	R4
	1.543		0.416		57.692	S4
	1.473		0.070		57.762	T4
	1.547			0.074	57.688	U4
	2.268			0.721	56.967	P3
	2.200		0.068		57.035	Q3
	2.313			0.113	56.922	R3
	1.788		0.525		57.447	S3
	1.650		0.138		57.585	T3
	1.885			0.235	57.350	U3
1.738		2.780		0.895	56.455	TBM 'X' (CP4)
	1.565		0.173		56.628	P2
	1.490		0.075		56.703	Q2
	1.575			0.085	56.618	R2
	1.093		0.482		57.100	S2
	0.965		0.128		57.228	T2
	1.090			0.125	57.103	U2
	1.775			0.685	56.418	P1
	1.773		0.002		56.420	Q1
	1.778			0.005	56.415	R1/R0
	1.650		0.128		56.543	S1
	1.282		0.368		56.911	T1
	1.480			0.198	56.713	U1
	1.778			0.298	56.415	R0/R1
	1.708		0.070		56.485	S0
	1.575		0.133		56.618	T0
1.388		0.871	0.704		57.322	CP5
1.358		1.269	0.119		57.441	CP6
		0.994	0.364		57.805	OBM 42 Filbert St. East (57.80)
7.569	8.994	4.188	5.613	59.230	misclosure – 5 mm
-8.994			-5.613		-1.425	
-1.425			-1.425			

Surveyor D. Powey

Reduced by D. Powey

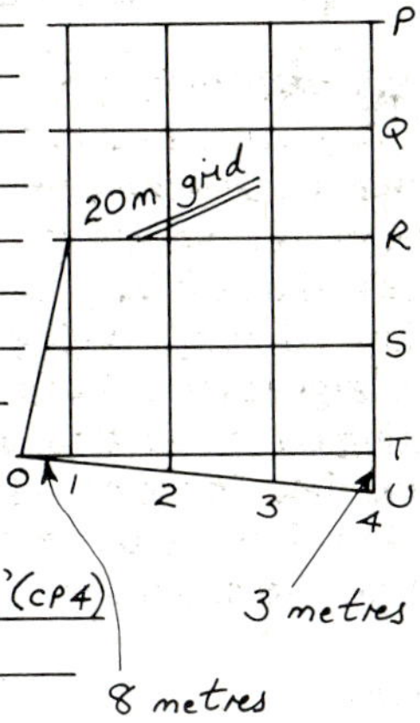

20m grid

3 metres

8 metres

Figure 5.46

In practice, it may on occasion be necessary to determine the average height of a site or area, as in earthwork calculations. An approximate value of the average height of the site may be obtained by simply dividing the sum of the heights of all the grid points by the number of these 'spot heights' involved, but this will give a poor result unless the site is reasonably level. A better approach is to find the mean height of each grid square (or other plan shape) using the values

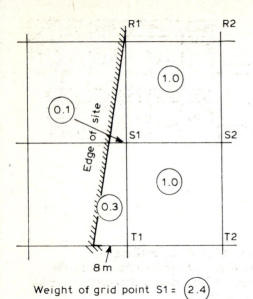

Weight of grid point S1 = (2.4)

Figure 5.47

'Weight' refers to the number of grid squares (or part grid squares) in which the reduced level appears.

5.6.3 LEVELLING FOR SECTIONS AND CROSS-SECTIONS

In building and engineering work it is often necessary to prepare a 'profile' of the ground along a particular line. Such a profile, termed a 'section of levels' or 'longitudinal section', is obtained by running a line of levels along the required line. Typical applications would be for roads, drain and sewer lines, etc.

Where the proposed construction is of considerable width, for example in roadworks, the longitudinal section information must be supplemented by 'cross-sections'. A *cross-section* is a profile of the ground, taken at right angles to a *longitudinal section* (*Figure 5.48*). The plotting

obtained for the grid points at the corners of the squares, then arrive at a mean height of all the squares. The method is illustrated in *Figure 5.47* and the Table following, which are based on the field observations in *Figure 5.45*. In the Table,

Grid point	RL	Weight	RL × weight
P4	57.28	1	57.28
Q4	57.35	2	114.70
R4	57.30	2	114.60
S4	57.69	2	115.38
T4	57.76	1.1	63.54
U4	57.69	0.1	5.77
P3	56.97	2	113.94
Q3	57.04	4	228.16
R3	56.92	4	227.68
S3	57.45	4	229.80
T3	57.58	2.2	126.68
U3	57.35	0.2	11.47
P2	56.63	2	113.26
Q2	56.70	4	226.80
R2	56.62	4	226.48
S2	57.10	4	228.40
T2	57.23	2.1	120.18
U2	57.10	0.1	5.71
P1	56.42	1	56.42
Q1	56.42	2	112.84
R1	56.42	2.1	118.48
S1	56.54	2.4	135.70
T1	56.91	1.3	73.98
U1	56.71	0.1	5.67
S0	56.48	0.4	22.59
T0	56.62	0.3	16.99
Totals 26	1482.48	50.4	2872.50
Approx mean	= 57.02	mean	= 56.99

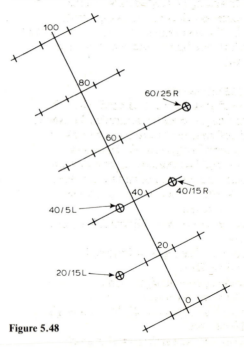

Figure 5.48

of such sections, together with the calculation of volumes from them, is described in *Site Surveying 3*.

On longitudinal sections, levels are observed generally at a standard interval of horizontal distance along the section line, typically 20 m, and at points where the ground changes significantly, or where natural or artificial features disturb the ground profile. On cross-sections, levels may be taken at standard intervals and always at natural and artificial features and significant changes of ground slope. Levelling for

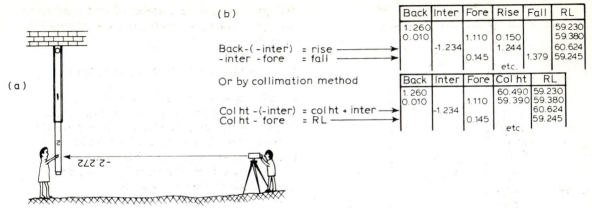

(b)

Back	Inter	Fore	Rise	Fall	RL
1.260					59.230
0.010		1.110	0.150		59.380
	-1.234			1.244	60.624
		0.145		1.379	59.245
				etc.	

Back - (-inter) = rise →
-inter - fore = fall →

Or by collimation method

Back	Inter	Fore	Col ht	RL
1.260			60.490	59.230
0.010		1.110	59.390	59.380
	-1.234			60.624
		0.145		59.245
			etc.	

Col ht - (-inter) = col ht + inter →
Col ht - fore = RL →

(a)

-2.272

Figure 5.49

longitudinal and cross-sections may be likened to a grid of levels which has been stretched in one direction and contracted in the other, hence the method of levelling is very similar to grid levelling. Cross-section points are generally described by the number of metres left or right of the centre line, the longitudinal section line (*Figure 5.48*).

5.6.4 INVERSE LEVELLING

Inverse levelling is the method used when the point whose height is to be determined is above the line of collimation of the level, as bridge soffits or in obtaining vertical clearances (*Figure 5.49(a)*). The readings taken with an inverted staff are booked as negative values (*Figure 5.49(b)*), but in other respects the reduction is as usual.

Similarly at changepoints (*Figure 5.50*), using rise-and-fall:

Backsight − (−Foresight) = Backsight + Foresight = Rise,

and (−backsight)−foresight = fall.

Using height-of-collimation:

Coll. ht.−(−foresight) = coll. ht. + foresight = red'd level, and

Red'd level + (−backsight) = red'd level− backsight = coll. ht.

If any doubt arises as to the arithmetic processes, draw a sketch as *Figure 5.50*.

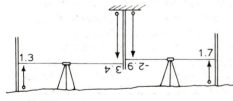

1.3

1.7

-2.9 3.4

Figure 5.50

5.7 Sources of error in levelling

Aim: *The student should be able to state the sources of induced and instrumental errors in levelling.*

Just as with linear measurement, levelling is never free from error. *Induced errors* are those attributable to the personnel involved, the nature of the ground or the climatic conditions, while *instrumental errors* are those which may be attributed to the equipment used. Some of these are due to carelessness, some have cumulative or constant or systematic effects on the results of the levelling.

The following section lists sources of errors and recommends precautions to be taken to eliminate them.

5.7.1 SOURCES OF INDUCED ERROR
(a) *Errors attributable to the surveyor:*

Mistakes in reading the staff
— read all three hairs.
Mistakes in booking readings
— book stadia readings in 'remarks' column.
Disturbing level and/or tripod
— check position of bubble; do not lean on or kick tripod.
Failure to level the bubble
— check before reading.
Incorrect focusing
— eliminate parallax.
Mistakes in reducing levels
— carry out mathematical checks.

(b) *Errors attributable to the staffholder:*

Not holding the staff upright
— check staff bubble.
At change points, not ensuring that the staff is

held in exactly the same position for both back- and foresights at a point
— use staff support.
Unequal back- and foresights
— surveyor check occasionally using stadia hairs.
Staff not properly extended
— surveyor check as needed by viewing connecting portion(s) of staff through telescope.

(c) *Errors attributable to the ground or climatic conditions:*

Sinking/rising of the instrument and/or staff
— set up on firm (not frozen) ground, use staff support.
Strong winds, staff and instrument vibrating
— find sheltered spot for instrument and staff, set tripod up low with its legs spread (*Figure 5.19*) and tread its feet well in (if automatic level, hold tripod lightly); brace staff with two ranging rods (*Figure 5.51*); if wind becomes too strong, cease work.
Heat shimmer, staff graduations unsteady (appear to bounce)
— reduce length of sights; try to keep line of sight well above ground level.

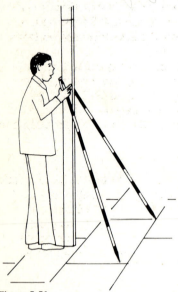

Figure 5.51

Direct heat of the sun causing differential expansion of instrument parts, bubble tube in particular:
— hold field book to shade bubble tube when levelling up (survey umbrellas can be bought).
Curvature and refraction (but no visible disturbance of image)
— these errors generally negligible, but reduce or eliminate by using equal sight lengths; keep sights short (max. 60 m); avoid grazing rays (sights near ground), avoid continually reading zero end of staff (e.g. when going uphill).

5.7.2 SOURCES OF INSTRUMENT ERROR
(a) *Errors attributable to the level:*

Faulty permanent adjustments
— check adjustments from time to time (see § 5.3.5); keep backsight and foresight lengths equal.

(b) *Errors attributable to the tripod:*

Play in the joints
— check occasionally; tighten as necessary.

(c) *Errors attributable to the staff:*

Longitudinal warping of the staff
— errors usually insignificant, but they are cumulative; if serious, discard staff.
Graduation errors
— an error in the 'zero point' of the staff has no effect, but errors at other points on staff may be cumulative; occasionally check graduations against steel tape.
Staff bubble out of adjustment
— check occasionally; error minimal but cumulative.

If level readings fail to close within the required tolerance, the only possible acceptable action is to re-level in the field.

The souces of error outlined above should be considered well in planning the re-levelling.

Part D — Angular measurement

6 The measurement of horizontal angles

Introduction
Angular measurement has been defined as the measurement of angles in the horizontal and vertical planes. Surveys may be carried out without angular measurement but it may be extremely useful, particularly in saving time, to have a piece of equipment capable of being used to obtain angles in either horizontal or vertical planes (or both). How to measure and make use of vertical angles generally is dealt with in *Site Surveying 3,* while this text considers the measurement (and to some extent the use) of horizontal angles using a theodolite. Horizontal angle-measuring equipment of limited value has been mentioned previously, particularly the prismatic compass, the optical square and the graduated circle fitted to some levels (§§ 2.1.4, 3.3 and 5.1.1).

The theodolite is an instrument designed specifically for the accurate measurement of horizontal and vertical angles required for the production of maps and plans, for setting out, and for the calculation of areas. It can be an extremely accurate piece of equipment, some instruments being capable of being read to the nearest minute of arc while others can be read to

0.1 of a second of arc. A theodolite measuring to 20˝ of arc is usually sufficient for site surveying. The basic principles of the instrument are relatively simple, and it is the most versatile of survey instruments, capable of being used for a wide variety of tasks. Applications include setting out lines and angles, levelling, optical distance measurement, plumbing tall buildings and deep shafts, and geographical position fixing from observations of the heavenly bodies (sun and stars).

6.1 The basic components of the theodolite
Aim: *The student should be able to describe the basic structure of the theodolite.*

Figure 6.1 shows the basic construction of early instruments first made about 400 years ago, and modern instruments are merely refined versions of this original concept.

A circular protractor, today called the *horizontal circle* or *lower plate,* was supported horizontally with its engraved surface lying uppermost. Above this plate was an *upper plate* with an index pointer, the plate being rotatable to

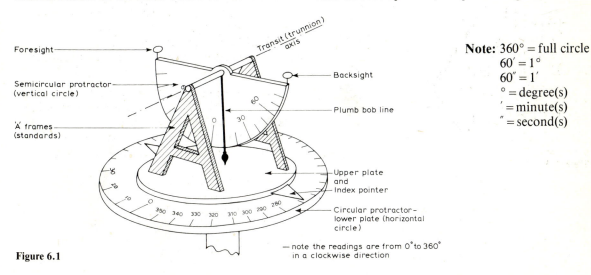

Note: 360° = full circle
60′ = 1°
60″ = 1′
° = degree(s)
′ = minute(s)
″ = second(s)

Figure 6.1

— note the readings are from 0° to 360° in a clockwise direction allow the pointer to be laid against any graduation on the lower plate. The upper plate carried two vertical A-frames (now known as *standards*) supporting between them a horizontal axis, known as the *transit or trunnion axis,* at the centre of which a semicircular protractor was rigidly fixed at right angles to it. A weighted plumb-line was also carried by the transit axis.

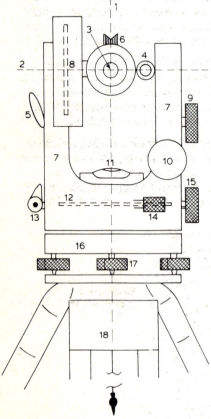

1 Vertical axis
2 Trunnion axis (at right angles to vertical axis)
3 Line of collimation (at right angles to trunnion axis)
4 Circle(s) reading eyepiece
5 Daylight reflecting mirror
6 Telescope and open sights
7 Standards
8 Vertical circle (within body of instrument)
9 Vertical clamp, and
10 Slow-motion screw
11 Horizontal plate bubble
12 Horizontal circle (within body of instrument)
13 Horizontal circle locking lever (or lower plate clamp)
14 Horizontal (upper plate) clamp, and
15 Slow-motion screw
16 Levelling head (tribrach)
17 Levelling screws (some theodolites have a ball-and-socket
   joint with levelling cams)
18 Tripod
19 Plumb-bob
Figure 6.2

(The modern protractor is a full circle and is termed the *vertical circle.*)

On its diameter the semicircle carried two sights, as on a rifle, which defined the line of sight, or line of collimation. To read at the instrument station a horizontal angle between two targets, the sights were aimed alternately at the left-hand and right-hand targets and the horizontal circle readings to them were noted, the difference between these readings giving the horizontal angle between the directions to the targets as measured at the instrument station.

To read a vertical angle, the sights were aimed at a target and the angle of elevation or depression read off the vertical circle against the weighted plumb-line.

Modern theodolites have telescopes to improve the line of sight in distance and clarity of definition. They also have glass circles within the body of the instrument, replacing the original external brass protractors, thus the circles may be graduated more finely and light can be reflected through the circles to show the markings with greater clarity. *Figure 6.2* shows the essential components of such a theodolite.

6.2 Types of theodolite
Aim: *The student should be able to describe the various types of theodolite.*

In the United Kingdom there are available some thirty different theodolites produced by a dozen different manufacturers. All of these may be of some use to the site surveyor, as all can be read directly to some value between one minute and one second.

Although the principles of modern theodolite construction are largely as outlined above, the ways in which manufacturers implement these principles can vary widely, and it is important to have an understanding of the main variations which may be encountered. *Figure 6.3* illustrates the following sections:

(a) *Tripod type:* The extending tripod as used on levels is still in general use for theodolites, but some versions are equipped with a centring rod (like a central fourth leg, but non-supporting) which has an attached circular bubble and is used for centring the instrument over the ground mark. Some models of these have a special levelling/centring head built into the top of the tripod, instead of the more usual flat tripod head.

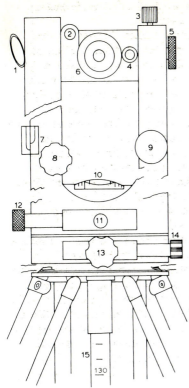

1 Daylight reflecting mirror
2 Collimator sight
3 Vertical clamp
4 Optical reading micrometer, and
5 Adjusting screw
6 Telescope eyepiece
7 Altitude bubble (co-incidence reader), and
8 Adjusting screw
9 Vertical slow-motion screw
10 Plate bubble
11 Upper plate clamp, and
12 Slow-motion screw
13 Lower plate clamp, and
14 Slow-motion screw
15 Centring tripod (centring rod)

Figure 6.3

(b) *One or two horizontal plates:* The existence of the traditional two-plate construction is indicated by the presence of two horizontal clamps and two slow-motion screws for the control of rotation about the vertical axis. If an instrument has only one clamp and slow-motion screw, then it is of single-plate construction. This type will have an additional control for horizontal movement, either a circle-locking lever (also termed a 'repetition clamp') or a circle-orientating gear knob protected by a hinged cover. These control arrangements affect the methods to be used in setting or reading horizontal angles.

(c) *One or two bubble tubes:* On its upper plate a

theodolite has a bubble tube which is used in the same way as the bubble on a dumpy for levelling up the instrument. If a second bubble is located on one of the standards, then this bubble must be centred before a vertical angle is read. The second (altitude) bubble will have its own adjusting screw additional to the clamp and slow-motion screw controlling movement of the telescope. Many modern theodolites are fitted with automatic indexing (self-zeroing) vertical circles, and there is no need to centre a second bubble with these types. (See *Site Surveying 3*).

(d) *Micrometer knob:* The presence of an additional knob or screw which is not used for movement control would suggest that the instrument is fitted with an optical micrometer reading system, described in § 6.2.1(c).

(e) *Externally visible circles:* If the graduated circles are not encased fully, and if they can be seen without using an eyepiece, then the instrument is probably an obsolescent vernier type (*Figure 6.4*). Modern instruments have

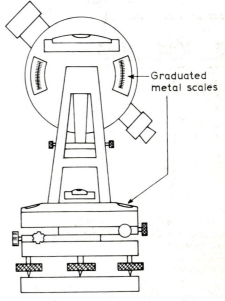

Graduated metal scales

Figure 6.4

completely enclosed glass circles which are read through an additional eyepiece close to the main telescope eyepiece. Glass-circle systems require illumination and this is usually provided by adjustable mirrors reflecting daylight into the instrument, or by fitting battery-powered lamps.

(f) *Collimator:* Some instruments have a very small tube, like a miniature eyepiece, attached to the top of the telescope. This is a collimator sighting device used instead of open sights for aiming the telescope.

(g) *Graduation patterns:* Theodolite circles used in the United Kingdom are generally graduated in various arrangements of degrees, minutes and seconds, and very occasionally in degrees and decimals of a degree. Theodolites graduated in the grade system (360° = 400ᵍ), as used in some parts of Europe, and others using the mil system (360° = 6400 mils), are also available.

(h) *Reading systems:* There are four common types of reading system in use, according to the accuracy demanded of the instrument and the manufacturer's design preferences. These are described in § 6.2.1.

(i) *Accuracy classifications:* Generally, the significant element in the choice of a theodolite is the accuracy attainable with the instrument, and this depends upon a combination of its quality and inherent accuracy and the techniques used for angle measurement. Although instruments may be classified in a variety of ways, they are usually classed primarily according to their accuracy, and the generally-accepted classification of theodolites is described in § 6.2.2.

6.2.1 THEODOLITE READING SYSTEMS
(a) *Circle microscope:* The most basic optical reading system having a simple fixed hair-line against which the circle graduations are read directly.

Example: The Kern KOS construction theodolite with circles graduated to five minutes of arc, thus allowing readings to be estimated to the nearest minute of arc. *Figure 6.5* illustrates a typical set of readings as viewed in the microscope. This instrument has three scales — the lowest being the horizontal angle, then the vertical, and finally a vertical gradient scale showing percentage gradients. (The last is not an item normally on theodolites.) The readings are:

Horizontal, 53°12′; vertical, 85°48′;
gradient, +7.35%

(b) *Optical scale:* Developed from the preceding type, this has a diaphragm fixed in the microscope, with a scale of divisions on it rather than a fixed hair-line.

Example: Figure 6.6 shows the Wild T16 scale-reading theodolite with the glass circle divided

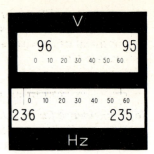

Figure 6.5   **Figure 6.6**

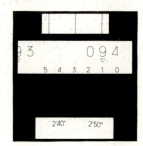

Figure 6.7   **Figure 6.8**

into single degrees only and the optical scale extending over one degree but divided to single minutes of arc. The number of the degree division is noted, then the number of minutes at the degree division, and finally the tens of seconds of arc are estimated (only one degree division can cut the optical scale unless the reading is an exact number of degrees). The readings are:

Horizontal, 235°56′20″; vertical, 96°06′30″

(c) *Optical micrometer:* This system combines the optical scale principle with lateral movement of the scale under the control of a micrometer setting screw which adjusts the position of a prism in the viewing system. The system is intended to provide greater accuracy than can be obtained with the simpler optical-scale type.

Example: The Wild T1 micrometer theodolite shown in *Figure 6.7* is typical. When the micrometer setting screw is turned, all the numbered graduations appear to move while the central single/double line remains stationary. The micrometer screw is turned until a degree division of the required scale (horizontal (Hz) or vertical (V)) is central between the fixed pair of lines, then the reading is the number of the degree division plus the minutes and seconds (or minutes and decimals of a minute in this example) shown in the third 'window'. It is important to note that a movement of the micrometer screw moves the *images* of the circles, it does not cause any movement of the circles themselves. The readings are:

Horizontal, 327°59.6′; vertical needs a re-setting of the micrometer.

(d) *Coincidence (or double-reading) optical micrometer:* This system is used on instruments in which the highest accuracy of reading is required, since it provides the mean of two readings taken at diametrically opposite points of the circle, thus minimising errors of eccentricity (errors arising from the axis of rotation not being completely concentric with the circle of graduations). The optical reading eyepiece presents graduations from both sides of the circle, and these must be placed equally on either side of a fixed hair-line or, alternatively, brought into coincidence using the micrometer setting screw.

Example: Figure 6.8 illustrates the reading system of the Wild T2 universal theodolite, the upper window showing the graduations brought into coincidence with one another, while the central window shows the degrees and tens of minutes and the lowest window shows minutes and seconds which can be read direct to one second. Usually a switch is used to change the circle being viewed.

The reading is: 94°12′44.4″

(e) *Vernier scale: Figure 6.9* illustrates a typical obsolescent vernier scale system used on instruments with metal circles. These have generally been replaced by the modern enclosed glass circle instruments.

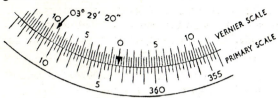

Figure 6.9 Primary scale at index mark 03° 20′ plus Vernier scale at 9′ 20″ gives a total of 03° 29′ 20″.

(f) *Digital electronic readouts:* These are used on some very modern instruments, at present primarily on specialised electromagnetic distance-measuring instruments.

6.2.2 THEODOLITE CLASSIFICATIONS
Theodolites are typically classed according to their reading accuracy, as indicated earlier, and

each class has a range of generally-accepted characteristics.

(a) *Precision theodolites:*
These types can be read directly to better than one second of arc and they are typically used by national mapping and land survey organisations. They are unlikely to be encountered by the site surveyor.

(b) *Universal theodolites:*
Also known as 'single-second' instruments, these allow readings directly to one second of arc. They are occasionally used on site survey work, generally when extreme angular accuracy is needed, such as possibly on very long sights.

Typical specification: Telescope magnification 30X; objective lens diameter 40 mm; direct reading on both circles to one second of arc (but a single reading is probably correct only to within ± 5″); coincidence optical micrometer; automatic vertical circle indexing; erect image; optical plummet and/or centring rod.

Example: Theo 010A, by Carl Zeiss, Jena, GDR (*Figure 6.10*).

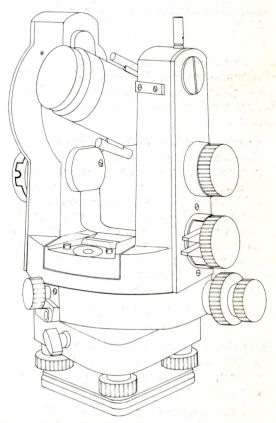

Figure 6.10 Zeiss Theo 010A universal

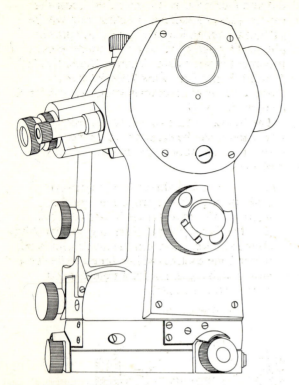

Figure 6.11 Kern K1-SE engineer's theodolite

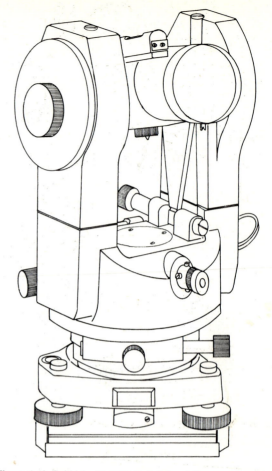

Figure 6.12 Sokkisha TM20E digital theodolite

(c) General purpose theodolites:
These read directly to about 20″ generally, and are fast and easy to use, ideal for general site survey work. Often described as 'engineer's' theodolites.

Typical specification: Magnification 25X to 30X; objective 35 to 45 mm diameter; direct reading to 20″ of arc or better; optical scale or optical micrometer reading system; automatic vertical circle indexing; erect image; optical plummet and/or centring tripod.

Examples: K1-SE engineer's theodolite, scale reading, by Kern, Aarau, Switzerland, and the TM20E 20″ digital theodolite by Sokkisha, Tokyo, Japan (*Figures 6.11* and *6.12* respectively).

(d) Builders' theodolites:
These instruments are of a comparatively low order of accuracy, reading direct to 1, 5 or 10 minutes of arc and by estimation to 30″ or 1′. They are usually rugged, simple to operate, and relatively cheap.

Typical specification: Magnification 15X to 20X; objective 25 to 35 mm diameter; circle

microscope reading system; no automatic vertical circle indexing; no optical plummet; erect image.

Example: Th51 minute-reading theodolite, by Carl Zeiss, Oberkochen, West Germany (manufactured in the USSR) (*Figure 6.13*).

Figure 6.14 illustrates diagrammatically the relative accuracies of universal, general purpose and builders' theodolites.

6.3 Field procedure — observing horizontal angles

Aim: *The student should be able to describe the procedures used to measure horizontal angles.*

Various methods of measuring horizontal angles have been used, and some are appropriate to older equipment (such as vernier theodolites) or to specialised tasks. For the site surveyor's

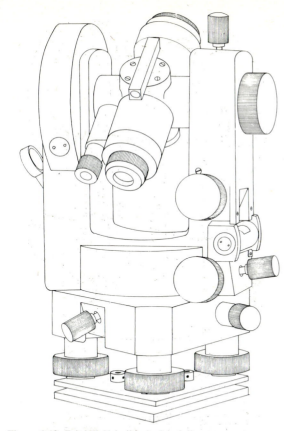

Figure 6.13 Zeiss Th51 builder's theodolite

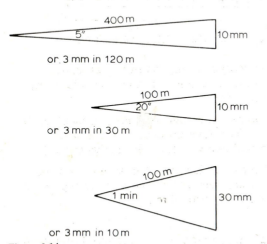

Figure 6.14

general work an observing technique is required which is easy to use, which provides some form of check on reading and instrumental errors and which will allow an increase in accuracy to be achieved readily when necessary.

A suitable method with modern equipment is that known as *simple reversal*, also called 'face

left and face right on one zero'. For most site survey work, measurement on one zero will be sufficient, but an increase in accuracy may be obtained by observing on two or more zeros. Building Research Establishment Digest No. 202, June 1977, *Site Use of the Theodolite and Surveyor's Level*, recommends the method for all construction work.

The following terms are used in theodolite work, and it will be necessary to define them before the observation routines can be fully explained.

Face left (FL) means that when the surveyor looks through the telescope eyepiece the instrument's vertical circle is located on his left. This is the normal observing position with all theodolites, and when the instrument is on face left the controls will be found to lie in the most suitable position for a right-handed observer (*Figure 6.15*).

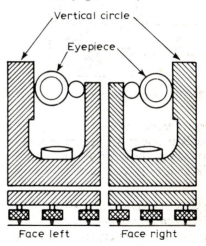

Figure 6.15

Face right (FR), accordingly, indicates that the vertical circle is on the observer's right-hand side as he looks through the telescope eyepiece.

Swinging, or *turning,* means rotating the instrument about its vertical axis. Accordingly, *swinging (turning) right* means rotating the instrument about its vertical axis in a clockwise direction in plan, while *swinging (turning) left* indicates that the instrument is being rotated in an anti-clockwise direction in plan.

Transitting the telescope means rotating it about its horizontal or *transit axis*, effectively reversing the direction in which the telescope is pointing. If, after a target has been observed on FL, the telescope is transitted and the instrument turned until the telescope again points on the target, the instrument will now be on FR.

If the horizontal circle is kept stationary throughout the operation, and the circle is read carefully on both pointings, then in theory the two circle readings will differ by exactly 180°. In practice, however, there is very often a small error, but it does provide a check on reading errors and instrument errors.

Reference object (RO), *back object* and *back station* are all terms used to denote the first target the theodolite is pointed at when measuring the horizontal angle subtended at a point or station by two distant target points or stations.

Forward object and *forward station* are terms used to denote the second target point observed on when measuring a horizontal angle.

A *single horizontal angle* measurement is made by pointing on the RO, booking the horizontal circle reading, then turning to point the forward object and again booking the circle reading, all on FL. The difference between the two circle readings is the required horizontal angle, and the reading when pointed on the reference object is the *zero* of the measurement. If, as an example, the circle read 45° when pointed on the RO and 65° when pointed on the forward object, then the horizontal angle is 20°, measured from a *zero* of 45°. In practice, the zero could actually be 0°; this is a matter of choice, as will be seen later.

Simple reversal means to measure a single horizontal angle as above, then transit the telescope onto FR, point the forward object again, read and book the circle, then point the RO and book the reading. This procedure gives two measures of the angle, the first on FL and the second on FR, the mean of the two values being accepted as the value of the angle if there is no gross error. Note that the pairs of readings (FL and FR) for each target direction provide checks on reading and instrumental errors. A value of an angle obtained in this way, with the graduated horizontal circle kept stationary in one position throughout the operation, is a value obtained by *simple reversal on one zero*.

The accuracy of this first value may be improved by moving the horizontal circle to a new zero setting, repeating the operation to obtain a second value, then taking the mean of the two values. The final value of the angle may be said to have been taken by *simple reversal on two zeros* which, barring gross errors, is more accurate than using only a single zero.

As a general rule, then, angles in site work should be measured by simple reversal, and for higher accuracy two or more zeros should be used and the set of results meaned. The reversal procedure cancels the effect of errors of eccentricity and collimation, the change of zero minimises the effect of circle graduation errors.

6.3.1 SETTING UP THE THEODOLITE
The method to be used depends upon the make and type of both the instrument and the tripod, there being three main variations of the latter.

(a) Simple tripod and plumb-bob
Set the tripod approximately over the ground mark, in the same way as for a level, with the top plate horizontal. If a small stone dropped from below the centre of the tripod head falls within 20 or 30 mm of the ground mark then tread the tripod feet well in, keeping its head horizontal. If the distance is greater, move the tripod feet equally laterally and repeat as needed, then finally attach the instrument to the tripod.

Attach the plumb-bob to the instrument, adjusting the cord length so that the bob clears the station mark (*Figure 6.16*). To *centre* the plumb-bob (and hence the vertical axis of the theodolite) vertically over the station mark, extend or shorten one or more of the telescopic tripod legs, as needed. In addition, the tribrach may be slid over the tripod head for fine adjustment, after easing off the clamping bolt (on some instruments the tribrach remains fixed and the body of the instrument above it may be slid to centre the plumb-bob). Raising or lowering the tripod's legs may tilt its top and make it impossible to level up the theodolite with the footscrews, yet the horizontal circle must be made horizontal before angles are measured. Since the sliding movement on the tribrach is limited, a compromise is ncessary between adjusting the legs and sliding the tribrach, but a considerable tilt of the tripod head must occur before it becomes impossible to get the instrument level with the footscrews.

Initial swinging of the plumb-bob should be damped by hand, but on windy days some form of screen may be needed. This can be provided by the surveyor crouching with his back to the wind,

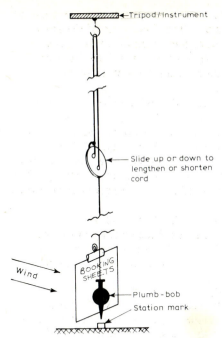

Tripod/Instrument

Slide up or down to lengthen or shorten cord

Wind

BOOKING SHEETS

Plumb-bob

Station mark

Figure 6.16

or by sheltering the bob with the field book or booking sheets.

When centring is complete, level up the instrument in the same way as described for the dumpy, using the plate bubble.

(b) Tripod and optical plummet
An optical plummet (*Figure 6.17*) is fitted to most modern theodolites, except for builder's types. This device gives greater centring accuracy, especially in windy conditions, but only if the instrument is level, and this may cause problems initially if the surveyor has used the plumb-bob technique to centre the theodolite before levelling it up.

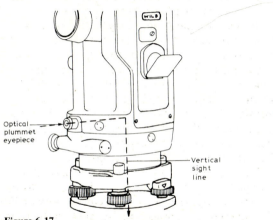

Optical plummet eyepiece

Vertical sight line

Figure 6.17

Having set up the tripod approximately over the ground mark, as described above, attach the instrument to the tripod. Centre the small circular bubble (generally fitted to the tribrach) using the footscrews, in the way described for the tilting level, thus roughly levelling the instrument.

Focus the optical plummet by first rotating its eyepiece to bring the cross-hairs (sometimes a target ring) into sharp focus then pushing in or pulling out the eyepiece to bring the ground into focus. If the ground mark is not visible, the tripod must be moved or its legs extended and/or shortened. The ground point vertically under the instrument can be found by moving a toe-cap around until it comes into view in the optical plummet, then marking that point will give an indication of which way the tripod must be moved (*Figure 6.18*). With the tripod re-adjusted,

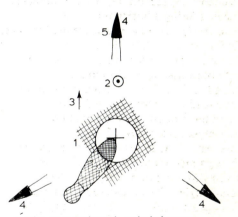

1 View as seen through optical plummet
2 Ground mark outside field of view
3 Direction of movement required by tripod
4 Tripod feet
5 Shortening this leg, or moving it away from the centre, will bring cross-hairs closer to the ground mark

Figure 6.18

the smaller circular bubble re-centred, and the ground mark visible, ease the tripod clamp bolt and slide the instrument (without turning it) until the ground mark coincides with the plummet cross-hairs.

On completion of centring, level up the instrument in the same way as described for the dumpy, and look through the optical plummet to check the centring, re-centring and re-levelling as necessary. On most instruments, the accuracy of the optical plummet may be checked by turning through 180° in plan and sighting the ground mark again — if the optical plummet is in adjustment, the points viewed before and after the 180° turn should coincide.

(c) Centring tripod

The centring tripod has the advantage that the site surveyor can centre extremely quickly (set-up time is halved) and, when centred, the instrument is approximately level. All Kern general purpose and universal theodolites are supplied with this facility, and some other makers provide it as an optional extra. In some survey tasks, an additional advantage is that the height of the instrument (to centre of the transit axis, usually) can be read directly off the centring rod (centre leg).

The centring tripod and its centring rod are telescopic, the retracted rod being held by a spring clip which is released by a sharp, light, downwards pull. The centring rod carries a small bubble. Most site surveyors who have used a centring tripod are reluctant to go back to the traditional type. *Figure 6.19* shows the Wild GST70 centring tripod.

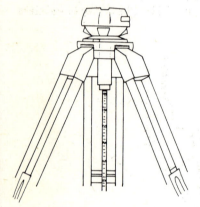

Figure 6.19

With the centring rod retracted, set the tripod approximately over the ground mark. Extend the rod, ensuring that the knurled clamping sleeve under the tripod head is unclamped, and place the point of the rod on the centre of the ground mark (it may be advisable to centre-punch the ground mark). Adjust the tripod leg lengths until the centring rod circular bubble is roughly central (floating). Make the final centring adjustment by sliding the tripod head over the top of the tripod, using the simple locking device generally provided to enable this final adjustment to be carried out. When the centring rod bubble is central, tighten the knurled clamping sleeve. The rod may be rotated about its vertical axis to check if the bubble is in correct adjustment. When centring is complete, attach the instrument to the tripod and level up in the same way as described for the dumpy. (**Note:** Some instruments have levelling *cams* instead of footscrews.)

6.3.2 SETTING A SPECIFIC READING ON THE HORIZONTAL CIRCLE

It is often necessary to point the instrument on a target with a specific reading already set on the horizontal circle, as for example when a simple reversal measurement is to be carried out with a specified zero set on the circle. The process of setting a specified reading on the circle is termed 'setting on an angle', and the method to be used depends upon the construction of the instrument.

(a) Instruments with upper and lower plates, each controlled by its own clamp and slow-motion screw

In these instruments (example, *Figure 6.3*), if only the *lower plate clamp* is applied then the horizontal circle is clamped to the levelling head and the upper part of the instrument may be rotated about its vertical axis. If the instrument is turned, the readings on the horizontal circle viewed through the optical reading eyepiece will alter. If, however, only the *upper plate* is clamped, and the lower plate released, then the horizontal circle is clamped to the upper plate and rotates with the telescope, and in this case the readings will not alter as the instrument is turned. When *both* plates are clamped it should not be possible to turn the instrument (except by applying excessive force).

Accordingly, to set on a particular angle, first apply the lower clamp, then turn the instrument until the horizontal circle reading approximates to the required reading, then apply the upper clamp.

If the instrument is a simple circle microscope or optical scale theodolite. (*Figures 6.5* and *6.6*), use the upper plate slow-motion screw to set the desired reading exactly then unclamp the lower plate. The theodolite may now be turned and the required reading will remain set in the field of view of the optical reading eyepiece, and the operation is complete.

If the instrument is an optical micrometer type, then use the micrometer setting screw (*Figure 6.3*) to set the minutes and seconds and then turn the upper plate slow-motion screw until the required degree value is exactly central between the pair of lines (*Figure 6.7*). The required angle is set and the operation is complete when the lower plate is unclamped.

It should be noted that some theodolites have a

milled rim fitted to the circle, this saving the operator from having to walk around the instrument looking for the required angle reading. The Wild T1 (*Figure 6.7*) is of this type.

(b) Instruments with a single horizontal plate clamp, slow-motion screw, and a circle locking lever

The circle locking lever replaces the lower plate clamp and slow-motion screw (*Figure 6.20*), thus

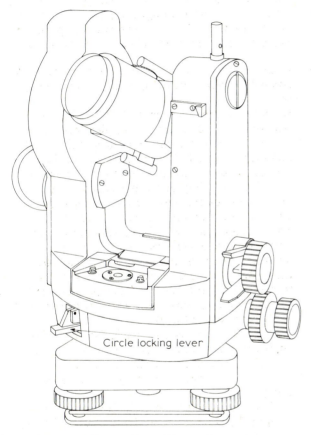

Figure 6.20 Zeiss 020A

the method is basically similar to that in (a) above. Various designs of locking lever are used, but in all cases they may be used either to lock the circle to the levelling head or to lock it to the horizontal plate.

To set on a particular angle, lock the circle to the levelling head using the circle locking lever, then release the horizontal plate clamp and turn the instrument until the horizontal circle reading approximates to the required reading, then apply the horizontal plate clamp. Use the horizontal slow-motion screw to set the desired reading

exactly, then use the locking lever to lock the circle to the hroizontal plate and release the horizontal plate clamp. The theodolite may now be turned and the required reading will remain set in the field of view of the optical reading eyepiece.

(c) Instruments with micrometer setting screws, circle orienting drive, and no lower plate

Since the horizontal circle of instruments of this type can only be moved by the circle orienting drive, and the circle cannot be swung freely as in the other types, it is not possible to set on an angle before pointing the target. Accordingly, before setting on the angle, the target must be bisected, as explained in § 6.3.3 (e).

To set on a particular angle, having pointed the target, use the micrometer setting screw to set the required minutes and seconds, then unclamp (or lift the protective cover over) the circle orienting drive and use it to set the required degree value (*Figure 6.21*). The drive should then

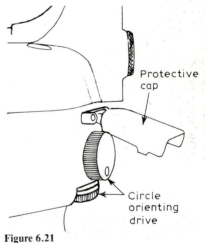

Figure 6.21

be re-locked (or the cover replaced) to prevent accidental movement of the circle. Note that in this case, if the horizontal clamp is released and the telescope is swung, the readings will not remain the same, but if the target is re-pointed the angle originally set should re-appear in the reading eyepiece.

6.3.3 OBSERVING THE DIRECTION TO A TARGET

The following procedure is used to point the telescope accurately at a distant target station and obtain the horizontal circle reading. It is *not* to be used when observing the RO with a specific zero value set on the circle (that procedure is set out in § 6.3.4).

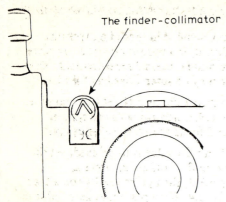

The finder-collimator

Figure 6.22

(a) Release the (upper) horizontal plate and telescope clamps. (The bracketed reference '(upper)' indicates that in the case of an instrument with two horizontal clamps, then it is the upper clamp which should be used.)

(b) Point the telescope towards the target and use the open sights (or collimator if fitted) to sight onto the target (The 'finder-collimator' of a Kern DKM2A is shown in *Figure 6.22*).

(c) Apply the (upper) horizontal plate and telescope clamps.

(d) Focus the target carefully, ensuring that there is no parallax.

(e) Bisect the target using the (upper) plate and telescope slow-motion screws. To do this, bring the target close to the centre of the cross-hairs with the slow-motion screws (*Figure 6.23*) then

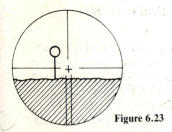

Figure 6.23

pause, take a breath and hold it, and slowly turn the horizontal slow-motion screw to bisect the target. The bisection achieved should be accepted and not 'fiddled with', it will improve with practice.

The target being bisected should preferably present a fine mark in the field of view of the telescope — a ranging rod can be seen thus at 1600 m (1 mile) or a matchstick at 100 m. For much construction work a good mark is provided by a small nail projecting 10 mm or so from a

wooden peg. The target must be vertical, and the cross-hair intersection should be aligned on the *base* of the target. Errors in bisection have a particularly significant effect on the accuracy of angle measurement when the sighting distance is short.

(f) Read and book the horizontal circle reading.

6.3.4 POINTING A TARGET WITH A SPECIFIC READING SET ON THE CIRCLE

This procedure is used when observing on the RO with a specific zero value set on the circle, as when making the first pointing on the RO in a simple reversal measurement.

For an instrument fitted with circle orientating drive, all that is required is to observe on the target as described in § 6.3.3, but without reading the circle, then set the required zero value on the circle as described in § 6.3.2, and the process is complete. The circle orientating drive should not be touched again until another zero setting is required.

For other types of instrument, the procedure is as follows:

(a) With the instrument on face left, set the required zero value on the horizontal circle as described in § 6.3.2.

(b) With the upper horizontal plate clamp applied and the lower plate clamp released (or the horizontal circle locked to the upper plate and the horizontal clamp released, for circle locking types) and the telescope unclamped, swing right to sight the target with the sights or collimator.

(c) Apply the (lower) horizontal plate and telescope clamps and focus the target carefully.

(d) Bisect the target as in § 6.3.3 but using the (lower) plate and telescope slow-motion screws.

(e) Read and book the horizontal circle reading. (This should, of course, be the same as the value originally set on the circle.)

(f) If the instrument is fitted with a circle locking lever, lock the circle to the levelling head and do not touch the lever again until a new zero setting is required. For other instrument types, do not touch the lower plate clamp and slow-motion screw again until a new zero setting is required.

6.3.5 CHANGING INSTRUMENT STATION OR PACKING UP

When a theodolite and tripod are being moved from one station point to another it is a common practice to leave the instrument attached to the tripod with its legs closed, the whole outfit being carried vertically to his front by the surveyor. It should be noted that there is always a danger of causing expensive damage by tripping up, or striking the instrument against a wall, etc. It is safer, particularly over a great distance, to carry the instrument packed in its carrying case.

All footscrews and slow-motion screws should be returned to the centre of their 'runs' before moving or packing the instrument. If carrying the instrument on the tripod, release the plate and telescope clamps. When packing the instrument, if it is the type that sits vertically in the case then usually the clamps sould be applied lightly, but if it is the type which 'lies down' then usually the clamps should be released.

A wet instrument should be wiped dry as far as possible before being placed in its case, since this is generally air-tight, and it should be removed from the case as soon as possible to allow it to dry out thoroughly. In some conditions, leaving a wet theodolite tightly sealed in its case may ruin it in one weekend.

When moving or packing-up, remember the 'bits and pieces', i.e. the tripod cap, the plumb-bob, carrying straps, instrument case, etc.

6.3.6 CHOOSING THE ZERO SETTING TO USE FOR A SIMPLE REVERSAL MEASUREMENT

The first horizontal circle reading which is booked in simple reversal is the reading obtained when pointing onto the reference object, this reading being the 'zero' of the measurement. Beginners often set $00°00'00''$ on the circle before pointing the RO, since this simplifies the arithmetic later. (Remember that the RO pointing reading must be deducted from the forward object pointing reading to get a first value of the horizontal angle.) This practice has the disadvantage that it takes longer to set $00°00'00''$ on the circle than any other value, and it is not recommended for very old theodolites because of the possibility of graduation errors and wear in the instrument.

A common alternative is to set a small angle (under 10°) on the circle before pointing the RO. This avoids the disadvantages of setting $00°00'00''$ yet means that the arithmetic is still not likely to be difficult.

Another widely-used method is not to set any particular value on the circle, but simply to point on the RO and book whatever the circle reading happens to be as the value of the first zero. This method is probably more appropriate to a more experienced operator of the instrument.

6.3.7 OBSERVING A HORIZONTAL ANGLE BY SIMPLE REVERSAL

The procedures detailed in preceding sections must be combined, in appropriate order, in order to carry out a simple reversal measurement. The whole routine may be summarised as follows:

(a) Set up the theodolite over the station point as described in § 6.3.1, according to the type of instrument, and focus the eyepiece carefully.

(b) If a specified zero value is to be used, set the required angle on the horizontal circle as described in § 6.3.2, then point the RO on FL as explained in § 6.3.4. If no particular zero value is to be used, then simply point the RO (as in § 6.3.3) and the reading which is booked will be the zero for the measurement.

(c) Point the forward object, again as in § 6.3.3.

(d) Release the (upper) plate and telescope clamps, transit the telescope, and swing left to point the forward object again, as in § 6.3.3 but on FR.

(e) Still on FR, swing left to point the RO, as in § 6.3.3.

(f) Calculate the mean value of the angle, as illustrated in § 6.4. Four circle readings should have been taken, these being: a reading on the RO, on FL, a reading on the forward object, on FL, a reading on the forward object, on FR, and a reading on the RO, on FR.

The difference between the first pair gives one value of the angle, the difference between the second pair gives another value, and the mean of the two values gives the final value of the measured angle.

(g) If desired, the whole process from (b) above may be repeated with a new zero value set on the circle in (b). If two zeros are to be used, the first could be made just over $00°00'00''$ and the second just over $90°00'00''$. If four zeros are to be used, then they could approximate to 0°, 45°, 90° and 135°, being just over in each case.

Two or more zeros are essential when an accuracy close to the limitations of the equipment is demanded. Thus, when using a one-second theodolite it is necessary to use four or more zeros to ensure that the mean angle is correct to within one second.

In addition, it is desirable to use two or more zeros if re-observation would entail a large expenditure of time and effort. In angular observation work generally a disproportionate amount of time is lost in getting to the site, setting up the equipment, etc, while the time spent on observing is minimal. An experienced site surveyor would have observed and booked the four readings above within ten minutes, but it would have taken as long to set up the equipment and still longer if targets for the reference and forward objects had to be established. Travelling time to the site must also be taken into account.

6.4 Booking horizontal angle readings

Aim: *The student should be able to record readings on the theodolite and calculate horizontal angles.*

As in chain survey and levelling booking, clarity and accuracy are essential in angle measurement booking. Field books or booking sheets ruled for angular measurement are not readily available, although large survey organisations and departments do provide these facilities. There are no international or even nationally accepted methods, but whatever method is used it must be comprehensible to another surveyor.

Figure 6.24 illustrates an example survey problem where a measured base has been run

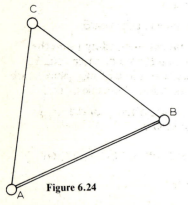

Figure 6.24

between two stations, A and B, and it is required to 'fix' a third point, C, by observing the horizontal angles of the figure at A, B and C. Assuming that FL and FR readings were observed on one zero (simple reversal using one zero) then one method of booking the readings is illustrated by the example booking sheet or field book page in *Figure 6.25*.

It should be noted that the FL and FR readings for pointings on the same target usually differ by 180° ± a small amount (§ 6.3). If, however, the FR minutes and seconds appear to be consistently lower (or consistently higher) than the FL values, it may be an indication of instrument error.

Instrument station	Face	Station observed	Horizontal plate readings	Reduced to RO	Accepted mean
A	L	RO Stn C	01° 04′ 20″		
		Stn B	62 26 40	61 22 20	
					61 22 20
	R	Stn B	242 27 00	61 22 20	
		RO Stn C	181 04 40		
B	L	RO Stn A	04 10 00		
		Stn C	61 32 40	57 22 40	
					57 22 10
	R	Stn C	241 32 00	57 21 40	
		RO Stn A	184 10 20		
C	L	RO Stn B	01 00 20		
		Stn A	62 16 00	61 15 40	
					61 16 00
	R	Stn A	242 15 40	61 16 20	
		RO Stn B	180 59 20		

Total	180 00 30
Misclosure	+ 30″
Observer	} T. Jens
Booker	
Date	6th May 1980

Figure 6.25

The heading 'Reduced to RO' gives the equivalent included angle at the instrument station, that is to say the reading on the forward pointing minus the reading on the RO.

An overall check on the consistency of the work is given, in the case of the triangular figure here, by summing the angles and noting the difference between the sum and 180°, since the angles of a plane triangle should sum to 180°.

If the readings were to be observed on more than one zero, then they might be booked as in *Figure 6.26* which illustrates the measurement of

Instrument station	Face	Station observed	Horizontal plate readings	Reduced to RO	Accepted mean
A	L	RO Stn C	01° 04′ 20″		
		Stn B	62 26 40	61 22 20	
	R	Stn B	242 27 00	61 22 20	
		RO Stn C	181 04 40		
					61 22 10
	L	RO Stn C	90 20 00		
		Stn B	151 42 40	61 22 40	
	R	Stn B	331 41 40	61 21 20	
		RO Stn C	270 20 20		

Figure 6.26

the angle at A on two zeros — simple reversal using a zero of 01°04′20″, then simple reversal again but using a zero of 90°20′00″.

100

Part E — Building surveys

7 The measured survey of a small building and its plot

Introduction

One of the objects of site surveying is the production of a map, plan and/or a section which may be used in the planning of new works or developments. The site of such works may contain (or may be) a building or buildings. In the case of buildings, it is often necessary to prepare not merely a ground plan but also a plan of each floor in the building, together with measured *elevations* (external views) and *sections* ('cut-away' views) of the buildings. While in land survey the site surveyor may have to describe boundary features, vegetation and street furniture (pillarboxes, etc.), in the case of buildings he may have to describe constructional details and building services.

The essential difference between these surveys and the chain surveys described in Part B lies in the plotting scales used, and reference may be made to BS 1192, *Recommendations for Building Drawing Practice*. The plotting scales in chain survey are typically 1:1000 or 1:500, and in building surveys 1:100 or 1:50. The use of larger plotting scales is due to 'the need to communicate adequately and accurately the character and size of the subject'. As this may demand measurements accurate to the nearest 10 mm (and occasionally 5 mm) the chain is not suitable for taking the measurements, and offsets are unacceptable unless over very short distances of a metre or less.

It is not difficult to keep lines straight generally, since the lengths of the lines to be measured are often under one tape length. However, the network of lines required for the framework and the techniques used are very similar to chain survey, except in the booking. Usually all the measurements of one floor of a small building may be on one sheet of paper, so that the detail surveyed from all the lines may readily be cross-referenced for ease of plotting.

7.1 Equipment for building surveys

Aim: *The student should be able to identify the instruments used for taking internal and external dimensions.*

7.1.1 EQUIPMENT FOR MEASURING THE BUILDING
(a) 30-m synthetic or steel tape. Recommended tapes conforming to BS 4484 Pt. 1: 1969 are illustrated in *Figure 7.1*.

(b) 2-m retractable tape rule, as recommended by the same BS (*Figure 7.2*).

(c) 2-m folding rod.

(d) A4 clipboard, paper, pen, pencil.

(e) Ancillary equipment, depending upon the task, may include hand-lamp or torch, chalk, manhole keys, duster or piece of rag, plumb-bob and cord, sectional ladder, binoculars, etc.

(f) A level and staff are often required, and occasionally a theodolite.

7.1.2 EQUIPMENT FOR MEASURING THE PLOT DIMENSIONS
As for chain survey.

7.1.3 PLOTTING EQUIPMENT
See § 7.4.

7.2 Field procedure on the building site

Aim: *The student should be able to measure and record a building and its site*

7.2.1 PRELIMINARIES
(a) Check the exact purpose of the survey, e.g. simple site plan, site and building plan with drainage layout, full building survey, etc.

(b) Arrange access to the site and building and all permissions.

(a)

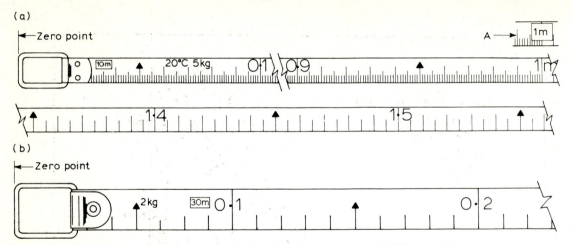

Figure 7.1 (a) Steel tapes. A shows an alternative marking for etched steel tapes. (b) PVC-coated fibreglass tape

(c) Check equipment, standardise tapes as needed.

(d) Obtain a sketch map of the site, or an outdated plan, or even a plan at a smaller scale, to provide a rough guide in reconnaissance and planning the line layout.

7.2.2 THE SITE RECONNAISSANCE

A reconnaissance must be carried out, in accordance with § 3.5, to decide upon the positions of the lines. Essentially, the lines must break the area into triangles, and each triangle must have a check, the triangles tied to one long base line through the site if at all possible. The base line is of particular importance on long, narrow sites, and it may be advisable to consider whether to use a theodolite to establish the line, especially if the ground is well covered with vegetation. Offsets will generally be few since it is often possible to run lines along straight boundary features. On occasion it may be found necessary to run one or more lines through the building(s), and it will be helpful if such a line can be arranged parallel to (or at right angles to) building faces or walls, since the interior of the building also is to be measured. *Figure 7.3* illustrates a typical plot with the lines it is proposed to measure.

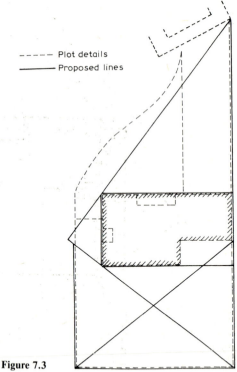

Figure 7.3

7.2.3 LINEAR MEASUREMENT

For preference, a steel tape should be used for the lines and a synthetic tape for the offsets, as

Figure 7.2

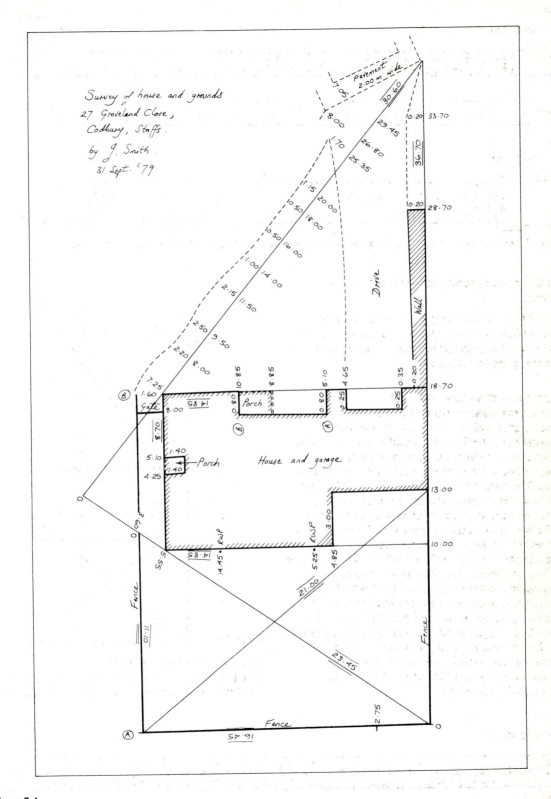

Figure 7.4

described in § 2.2.4. Sections 2.3 to 2.7 explain how to overcome typical problems arising during measurement.

7.2.4 BOOKING THE MEASUREMENTS

When the site is small and offsets are few in number, as is often the case, the booking may be carried out on a single sheet of paper, with single (skeleton) lines representing the chain lines. If the site is large, with possibly many offsets, it may be better to book the lines in the conventional chain survey manner (§ 3.2). The general guidelines of § 3.2 are relevant, whichever style is used, particularly with regard to accuracy and clarity.

Figure 7.4 illustrates such a page of booking. Note that the techniques of chainage, offsets and straights are similar, including the use of running measurements and lengths of lines to maintain accuracy and avoid ambiguity.

7.3 Field procedure — measuring the building

Aim: *The student should be able to explain how running internal and external measurements are taken, horizontally and vertically.*

7.3.1 PRELIMINARIES

(a) Check exactly what information is to be provided by the survey.

(b) Obtain keys to empty building(s), and any entry permissions required.

(c) Check equipment — tapes, folding rods, etc.

(d) Obtain existing plans, sections, elevations, if any.

7.3.2 THE BUILDING(S) RECONNAISSANCE

Carry out a thorough 'recce' of the building, to assess the size and nature of the task and decide how best to tackle it. A building is usually a series of blocks, of different shapes and sizes and at different levels, and the inexperienced surveyor must avoid the temptation to measure the blocks separately then fit them together like a '3D' jig-saw puzzle. The principles of survey of § 1.3 must be remembered, in particular the idea of working 'from the whole to the part' and the need for the independent check. However, it is always difficult to avoid the fragmentation of the measurements and where this occurs reference to the control framework must be made.

It is often difficult to build up triangles, but every possible diagonal measurement should be taken within rooms, and every wall must be measured, and taken together these can form the required triangular frameworks. It must never be assumed that a building or a room is rectangular, and the fact that two diagonals of a notionally rectangular room are equal does not mean that the room is rectangular. Again, walls may vary in thickness throughout their length, particularly at intersections with cross walls and partitions, and often such changes are detected only by very careful measurement and checking. When dealing with large, complex and irregularly-shaped buildings it may be advisable to provide a theodolite control traverse, as described in *Site Surveying 3*.

7.3.3 SKETCHES

Since the building shape is generally small and approximately rectangular on plan, the lines to be measured are usually run along the features to be surveyed, and the sketches are usually drawn first, the bookings (measurement notes) being added later. (This is the opposite of the methods used in chain survey.) These freehand sketches should be as large as possible, preferably on A4 plain or 5 mm square ruled paper, and roughly true to shape, but no attempt should be made to sketch to scale — the most important consideration is to ensure that all details are clear and the required measurements can be shown without crowding or ambiguity.

(a)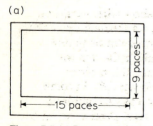

(b)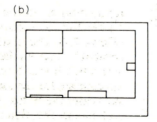

(c)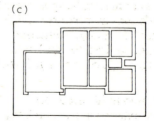

Figure 7.5 (a) Pace out the length and depth of the building to be surveyed — draw guide lines. (b) Sketch within the guide lines the outline of the building. (c) Draw the interior faces of the walls

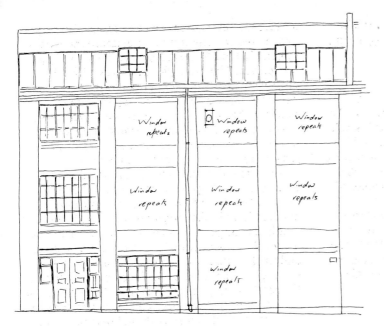

Figure 7.6

The following sketches are required:

(a) *Plans* of each floor of the building, including any basement, and possibly the plan within the roof space and/or a plan of the roof. A typical approach is shown in *Figure 7.5* but it should be noted that the external shape and size may already be available, in some cases, from field measurements and bookings carried out on the site survey.

Floor plans are drawn in the same way as external sites, that is to say looking down on the floor being drawn, but as if the building had been sliced through horizontally at about 1 m above the level of the floor concerned.

It may be possible to trace the ground floor plan sketch as a basis of the detailed sketches for other floors.

Details to be shown include wall openings (doors, windows, hatches, etc.), changes in wall thickness, changes in floor levels (steps, stairs, ramps, etc.) and similar constructional detail, with information as to materials, span directions, etc. Services installations and fixtures and fittings (gas, heating, water, electricity, ventilation, power, lighting, communication, cooking, sanitation, washing, etc.) and drainage layouts are all required.

(b) *Elevations* are used to illustrate the arrangement of the exterior of the building, hence sketches are usually required for all elevations of the building. These sketches need not be completed in their entirety where detail is repetitive (*Figure 7.6*).

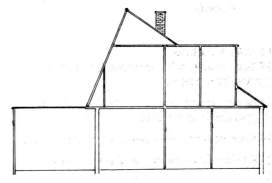

Figure 7.7

Detail to be shown is as for plans, but includes sills and lintels, external piping, and any detail which is visible on elevation but cannot be shown on plans.

(c) *Sections* are drawn to show what would be visible if the building was cut through vertically (*Figure 7.7*). The section need not be in one straight line across the building in plan (or vertically), it may be stepped if it is more appropriate for the job (*Figure 7.8*). In some such cases the section is actually a combined section and elevation in effect. Details shown on sections will be similar to those shown on plans and elevations.

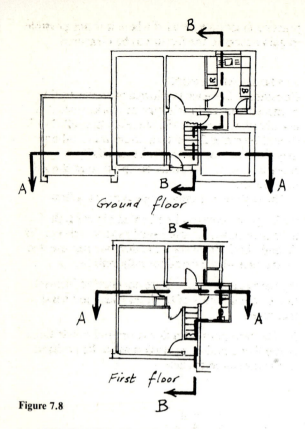

Ground floor

First floor

Figure 7.8

7.3.4 SYSTEM USED FOR MEASURING HORIZONTAL AND VERTICAL DISTANCES

External linear measurement can be carried out at ground level, as described for chaining in § 2.2.4, but due to the need to read the tape with a greater accuracy than is demanded in chaining, taping is often carried out with the tape held at chest height and making use of the hook attached to the zero ring of the tape (*Figure 7.9*).

It is conventional to take all measurements from left to right, i.e. anti-clockwise around the outside of the building and clockwise around the interior of rooms, since in this way the figures on the tape will be found to be the right way up and reading errors will be minimised. Also, when

plotting later from the bookings this method reduces possible misinterpretations as to which reading is which.

For ease and accuracy in plotting, all measurements along one straight line should be booked as one set of running measurements, provided that the accuracy of measurement can be maintained.

External vertical distances must be related to a selected suitable datum surface on the building, such as the line of a visible damp-proof course, a plinth or string course, etc.

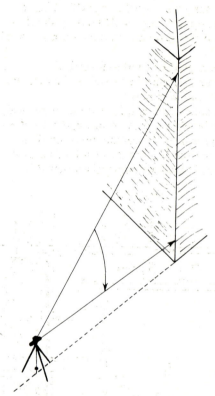

Figure 7.10 The theodolite is set up on approximate alignment of the building the upper corner of which is bisected through the telescope. Depressing the telescope should run its cross-hairs down the corner of the building, if this is vertical

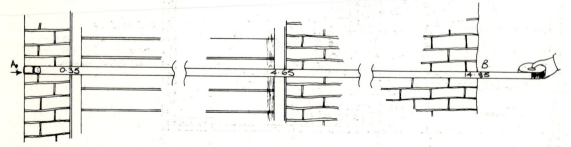

Figure 7.9 The tape is laid from left to right, and at chest height for ease of reading. Its hook at the zero end, A, grips the wall, while at B it is held with minimum tension to hold the zero end in place

If there is any doubt as to the verticality of a building face, it should be checked by theodolite. To do this, set the instrument up on the approximate alignment of the face and at about 5 to 10 m from the corner of the face, sight on and bisect the top corner of the face with the cross-hairs, then depress the telescope. If the face is vertical, the cross-hairs should appear to travel vertically down the corner of the building (*Figure 7.10*).

Where it is considered that the selected datum surface is not horizontal, this may be checked using the surveyor's level and staff. When the level is set up, the same staff reading should be obtained from all points on the datum surface if it is level.

Wherever possible, internal vertical measurements should be linked to the external heights at wall openings.

Inaccessible distances, horizontal or vertical, can often be obtained by counting the number of brick lengths or brick courses involved then directly measuring a similar length of accessible brickwork. It must be emphasised that this practice must only be used where it is not possible to gain access to the feature to be measured.

7.3.5 BOOKING METHODS

Considerable licence is allowed in the recording of measurements in this type of work, since it is usually plotted by the site surveyor himself, but it should be remembered that on occasion others may have to plot the work, hence clarity is essential and the generally accepted rules should be followed.

The methods usually adopted are as follows:

(a) Single or skeleton line booking is used, the lines themselves are not usually shown (except for tie and check lines) since the lines are represented by the face of the detail (usually wall faces).

(b) Letters and numbers are not used to identify the terminal points of the lines which are being measured.

(c) The zero point of a line is entered as 0 if there is any likelihood of confusion as to the position of the measurement zero.

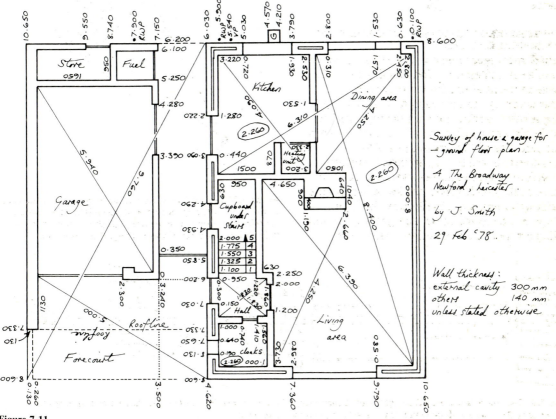

Figure 7.11

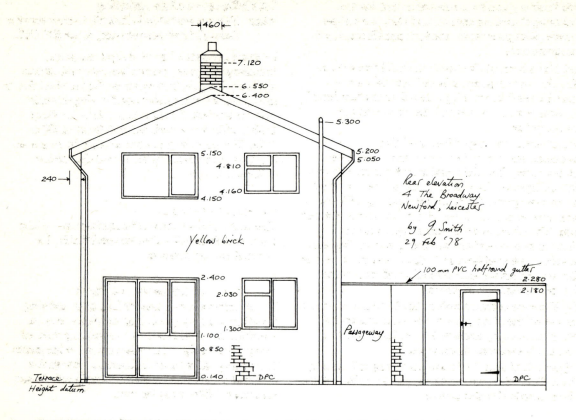

460

---- 7.120

---- 6.550
---- 6.400

— 5.300

5.150
4.810
4.160
4.150

5.200
5.050

240

Yellow brick

Rear elevation
4 The Broadway
Newford, Leicester

by J. Smith
29 Feb '78

100 mm PVC halfround gutter
2.280
2.180

2.400
2.030

1.300
1.100
0.850

Passageway

0.140   DPC

DPC

Terrace
Height datum

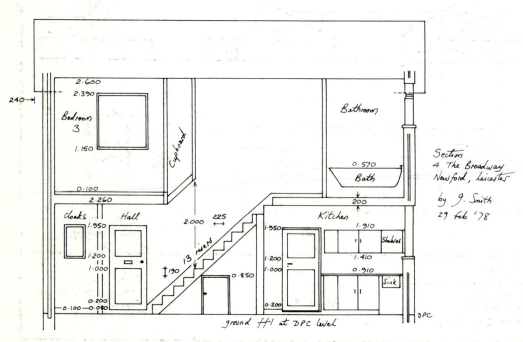

2.600
2.390

240

Bedroom
3

Bathroom

1.150

Cupboard

0.570   Bath

Section
4 The Broadway
Newford, Leicester

by J. Smith
29 Feb '78

0.100

2.260

200

Hall

2.000   225

Kitchen

cloaks

1.910

Shelves

1.950

1.950

1.200
1.000

13 risers

1.200
1.000

1.410

0.910

Sink

130

0.850

0.200

0.200
0.100 0.080

DPC

ground ffl at DPC level

Figure 7.12

(d) Running measurements are used for the 'chainage' measurements, and they are written, as previously recommended, in the direction of measurement.

(e) The length of the line is normally written as for chainage figures, it is only on the tie and check lines (i.e. the diagonals) that the line length is written with the base of the figure on the measured line.

(f) Offsets, running offsets and plus measurements may be written as in conventional chain survey booking, and may be entered either left or right of the line as space permits, if there is no ambiguity. Occasionally running offsets are also written as short 'chain' lines, but not tied out. To help minimise ambiguity, chainage figures may be written as metres and decimals of a metre, while offsets and plus measurements can be written simply as the number of millimetres, i.e. no decimal point shown.

(g) Floor-to-ceiling heights should be noted in the centre of the floor plan of the area concerned, the figures being encircled to indicate that they are not horizontal dimensions.

Figures 7.11 and *7.12* show an example booking for a floor plan, an elevation and a section, using the methods detailed above.

7.4 Office procedure, plotting

Aim: *The student should be able to plot surveys from field measurements, using BS 1192.*

Plotting consists of several separate plots, including site plan, floor plans, elevations and sections, and, depending upon the finished size of these, one or more sheets of drawing material may be needed. The site plan will often be at 1:500, while the remainder will be at 1:100 cr 1:50. The notes in § 3.7 concerning chain survey plotting are relevant here, in addition to the following points.

7.4.1 CHOICE OF SCALE
The scales chosen should be appropriate to the task, and these will often be specified in the original job description.

7.4.2. CHOICE OF DRAWING MATERIAL
For the alteration or conversion of an existing building, a heavy grade tracing paper may be suitable. Some individuals, however, prefer to carry out all original plotting on cartridge paper. If a strong, durable and dimensionally-stable material is necessary, a modern polyester-based translucent film must be used.

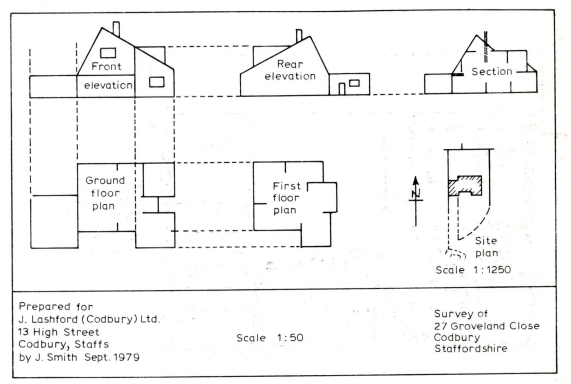

Figure 7.13

7.4.3 CHOICE OF SHEET SIZE

A measured survey of a small building and its site should generally plot on a single sheet of one of the recommended A-size drawing sheets, possibly A2 or A1, without being cramped.

7.4.4 LAYOUT OF THE SURVEY ON THE DRAWING MATERIAL

The ground floor plan is generally placed at the bottom lefthand corner of the sheet, with the front of the building towards the bottom of the sheet. The first floor plan is then placed alongside the ground floor plan with the same aspect or orientation, and the other floor plans similarly.

The front elevation of the building should be drawn immediately above the ground floor plan, again the other elevations are then drawn across the sheet in a row and level with the front elevation.

Finally, the sections and site plan should be placed wherever they will conveniently fit into the sheet arrangement, although the sections should, for preference, be placed alongside the elevations.

Wherever possible, it is important to arrange that common lines on plans, sections and elevations should lie on extensions of their respective lines on the drawing (*Figure 7.13*).

The presentation described above conforms closely to the 'first angle (or European) projection', and it is the projection recommended in BS 1192 to be used wherever practicable. Variations are inevitable, thus for example a small single storey extension at the rear of a building might not require a front elevation to be shown, and in each case the surveyor must decide the layout applicable in the circumstances.

In all cases the site plan should be drawn with north towards the top of the sheet.

Marginal and border information is mentioned in § 3.7, and this is relevant here. BS 1192 contains recommendations regarding these items also.

Final consideration regarding the layout is that the whole presentation should be well-balanced and pleasing to the eye, as with all drawings.

7.4.5 PLOTTING THE PLANS

The site plan is plotted like a chain survey, as covered in § 3.7.

The building itself is plotted in a similar manner, but the process is rather more complex.

The procedure is as follows:

(a) Draw the longest external wall to scale, in the preferred location on the drawing material.

(b) Using the measured wall thickness, plot the *alignment* (but not the detail) of the interior face of the same wall. Remember the possibility of different thicknesses along the length of the wall, and make allowance as needed.

(c) On the external wall face, plot the position of all wall openings, and raise right angles from these to cut the alignment of the inner face of the wall. Note that there may be rebated jambs, splayed jambs, and quite possibly the construction on either side of an opening may differ, and due allowance must be made for these details.

(d) Using the internal wall face measurements, scale off and mark the positions of all walls joining into the external wall.

(e) Using the conventional chain survey technique of plotting triangles by swinging arcs, plot the lines representing the faces of all the internal walls of the rooms adjoining the external wall already plotted, using wall lengths and room diagonals to build up the triangles.

(f) Repeat the process until all the ground floor walls have been plotted, applying all possible checks (matching internal and external measurements at openings, checking overall external wall lengths, etc.) since errors can build up rapidly when a building is plotted in this manner.

(g) When the ground floor plan framework is complete, and considered to be correct, plot all the detail involved.

(h) Plot the first floor plan by using the ground floor external dimensions (adjusted where necessary) and projecting across from the ground floor plan by the use of T-square and set square. Alternatively, trace the outline of the ground floor plan carefully, then locate the tracing paper where the first floor plan is to be plotted and 'prick' through the corners of the building onto the drawing material. The pricker marks may be joined in pencil to form the outline of the floor plan.

(i) Plot the interior of the first floor plan, in a manner similar to that used for the ground floor plan.

(j) Plot the remaining floor plans in the same

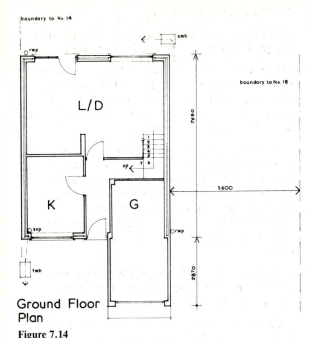

boundary to No. 14

L/D

K

G

Ground Floor Plan

Figure 7.14

way, adapt as needed if a roof plan is to be drawn.

Figure 7.14 illustrates a ground floor plot, being part of the drawing for an extension to a building.

7.4.6 PLOTTING THE ELEVATIONS

The front elevation should be plotted as follows:

(a) Project the lines of the external walls and openings of the front of the building upwards from the fround floor plan, as in *Figure 7.13*.

(b) Select an appropriate location and draw a line to represent the chosen datum line for heights, cutting across the lines projected up from the ground floor plan.

(c) Plot all the measured heights on the elevation, above or below the datum line as required, then complete the elevation drawing to show all the field sketch detail.

(d) Check the elevation, correct as necessary.

The remaining elevations should be plotted by a similar combination of projecting, tracing and scaling as practicable.

7.4.7 PLOTTING THE SECTIONS

For preference, these should be plotted alongside the elevations, allowing heights to be projected from elevations to sections and reducing the amount of scaling required. The sections are basically plotted in a similar manner to the elevations.

Figure 7.15 illustrates a typical elevation and section, again merely part of a whole drawing.

7.4.8 PENNING-IN — COMPLETING THE DRAWING

Section 3.7.4 deals with the completion of a chain survey plot, but it is equally relevant to building surveys.

Significant differences, however, include a probably greater use of graphical (point and area) symbols, and some of the examples from BS 1192 are shown in *Figure 7.16*. There will also be a greater use of descriptive names, abbreviations, annotations and numbers, and it may be necessary to show certain horizontal dimensions

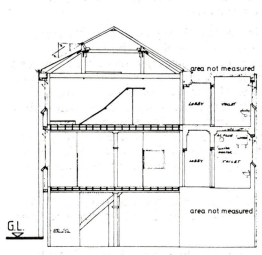

Figure 7.15   Section A—A

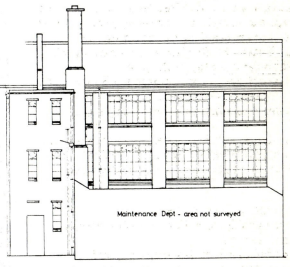

Rear elevation

and some heights. In view of these variations, a smaller type may be necessary for annotation generally.

Since at least two different scales may appear on the drawing, it is important that the user be able to readily identify the scale used on each part of the drawing. Scale lines are not usually shown on measured surveys of buildings. Again, two north points may be needed, to show the respective orientations of the site plan and of the ground floor plan.

Two line widths are often adequate for building surveys, as in *Figures 7.14* and *7.15*. *Figure 7.8* shows the planes of the sections passing through the plans, the letters identifying the section and the arrowheads the direction of view. This line should be shown using the broader gauge of nib and drawn as a broken line, as in *Figure 7.8*.

The method of showing stairs illustrated in *Figure 7.14* should be noted — each tread in a flight is numbered in succession 1, 2, 3 etc, commencing from the lowest tread, while the 'up' direction of the stairs is shown by an arrow and the word 'up'.

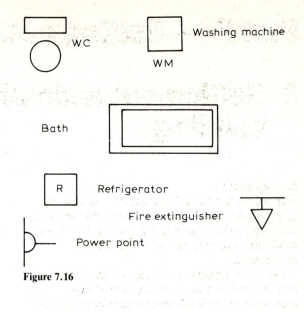

Figure 7.16

Finally, it must again be stressed that the object of the final presentation is to communicate information effectively to the user of the drawing.

Part F — Setting out

8 Setting out a simple rectangular building

Introduction

Setting out has been defined as the placing of pegs or other marks on or in the ground in such a way as to define the location and height of new works or developments which are to be constructed.

Only two aspects of setting out are considered here, these being the traditional methods employed for a simple rectangular building and for drainage work (§ 9.0). The control of complex buildings and the controls and calculations for roadworks are dealt with in *Site Surveying 3*.

The fundamentals of efficient setting out are considered to be accuracy, timeliness and clarity. In traditional building, setting out does not require high accuracy, but with the development of precisely-dimensioned frames and component members it is often found that insufficient attention has been paid to the initial site dimensioning and the resulting 'lack of fit' in panels, beams, etc., can be extremely expensive.

Building Research Establishment Digest No. 234, February 1980: *Accuracy in Setting Out*, states 'The degree of accuracy which can be achieved in construction has a significant bearing on design: in extreme cases, the feasibility of a structural system may depend on the standard of accuracy that can safely be assumed'.

Lack of accuracy is not only costly in rectifying mistakes, it may also be costly in time. If the site surveyor continually makes mistakes, the construction team will lose confidence and spend additional time carrying out their own checks.

Timeliness means not simply having the work set out on time, since 'time is money', but having the work set out early. This ideal is difficult to achieve because of, for example, possible vandalism of setting-out marks, unexpected changes in the construction programme, or the young site surveyor's uncertainty as to exactly what is expected. The advantages of being early are that the site surveyor does not need to work in haste or under pressure, conditions which can lead to undetected mistakes, and the construction team are provided with an incentive.

Clarity means that all personnel concerned should understand the surveyor's setting out plan and the meaning of all his pegs and other marks defining the location and height of the new works. Colour coding of pegs and other marks will assist in achieving this aim.

Adherence to the principles of survey covered in § 1.3 will assist in achieving these three fundamental aims.

8.1 Equipment

Aim: *The student should be able to identify and describe the equipment required, including a site square.*

The choice of equipment depends upon the nature of the particular job; for example, many small jobs can be set out without a theodolite. The following list includes the main items of equipment which may be required for setting out:

Theodolite: preferably general purpose with optical reading system (§ 6.2)
Surveyor's level: preferably automatic (§ 5.1)
Levelling staff: complete with hand bubble (§ 5.2)
Steel tape: 30 m (§ 2.1)
Ranging rods, arrows, optical square (§ 2.1)
Site square: an inexpensive optical instrument for setting out right angles (*Figure 8.1* and below)
Pens, pencils, field books, set square, scale, straight edge
Haversack for small equipment and sundries
Builders' line, various suitable hammers, cold chisel, centre punch, hand saw
Canoe level (*Figure 8.2*)
Wooden pegs: various sizes
Chalk, crayons, paints, paint brushes, nails, mild steel reinforcing rod, etc.
Profiles, profile boards, sight rails (*Figure 8.3* and below)

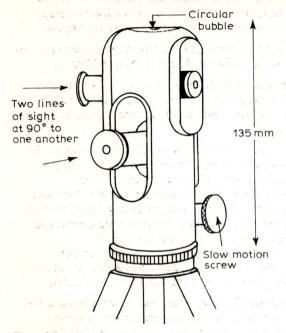

Figure 8.1

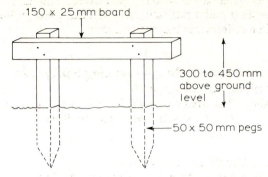

Figure 8.3 The length of the profile board depends upon the setting-out task, but it should be approximately 1 to 1.5 m. The length of the pegs is dependent upon ground conditions — minimum 600 mm

needed. They are used in conjunction with profiles/sight rails, *Figure 8.4* illustrating a typical traveller used to control excavation and foundation levels.

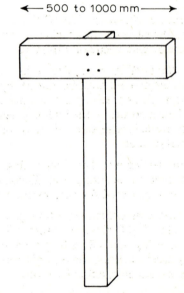

Figure 8.4 The length of the upright is to be calculated

Travellers, boning rods (*Figure 8.4*, §§ 8.4 and
 9.1, and below)
A *site square* (*Figure 8.1*) is a small instrument not used for survey work but only for setting out right angles in plan on construction sites in the absence of a theodolite. It consists simply of a cylinder carrying two mutually perpendicular horizontal axes, one above the other. Each axis carries a small fixed focus telescope which may be elevated or depressed but not turned horizontally. The telescopes then define two sight lines, at right angles to one another in plan, and if the whole instrument is turned until one telescope's cross-hairs bisect a reference object, the other telescope will define a line at right angles. The accuracy attainable is about 6 mm at a distance of 30 m.

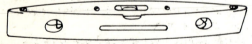

Figure 8.2

Profiles are arrangements of wooden boards and stakes or pegs, as will be explained later, used on site to define the alignments of wall and foundation faces, column positions, etc. Generally produced on site as needed, the form and dimensions vary with the task (*Figure 8.3*).

Travellers and *boning rods* are commercially available, but frequently are made on site as

Site markers, particularly for small jobs, are typically 50 mm square (or larger) wooden pegs or stakes, driven well into the ground, with a nail driven vertically part-way into the head of the peg. A fine nail projecting 10 mm or so above the top of the wooden peg provides a precise mark for sighting onto and for measuring to. If the ground is difficult, such as made-up hardcore of broken bricks, then it may be necessary to use short lengths of mild steel reinforcing rods as the site markers, with the exact point punchmarked on the end of the rod. In some setting out tasks it may be useful to cement a small steel plate in a required point position, then scribe a cross on the plate to indicate the precise point.

8.2 Field procedure — locating a proposed building on the ground

Aim: *The student should be able to explain how to set out a building and the likely constraints.*

8.2.1 STAGES IN SETTING OUT

The full procedure for setting out a building involves the following stages:—

(a) *Inspect the documents.* The relevant contract documents must be examined and checked for information relating to the setting out and any existing site survey data.

(b) *Inspect the site.* Just as for a survey, a full reconnaissance must be made before starting the job.

(c) *Prove the site drawings.* The 'fit' of the building on the site must be checked — errors in planning are often detected at this stage.

(d) *Set out the building plan.* The outline of the building on the ground must be marked by placing site markers precisely in the required position of each corner of the building, and sometimes on the intersections of cross walls, at column positions, etc.

(e) *Prove the setting out.* The final setting out must be checked for accuracy; e.g. diagonals calculated and measured for agreement, etc.

(f) *Set out initial height marks.* All heights on site should be related to Ordnance Datum, this generally requiring that TBMs be placed on site by levelling from OSBMs, then height reference pegs placed on site in suitable locations.

(g) *Set out profiles.* The site markers placed in (d) above define the building outline, but these marks will be destroyed when construction starts. Accordingly, the site markers must be used to set up profiles well clear of the works, these profiles then being marked with wall lines etc. in collaboration with the contractor. The contractor may then use the profiles to control line and height as the work proceeds.

8.2.2 INSPECTING THE DOCUMENTS

On large jobs special setting out drawings may be prepared for the site surveyor, but very often he must collect his data himself from the various documents produced during the job planning. The following documents may be relevant:—

(a) *Block plan:* This identifies the site, relates new work to existing. Scales 1:500 to 1:2500.

(b) *Site plan:* If this is to serve as a setting out drawing, it should be at a larger scale than the block plan, and include the crucial setting out data.

(c) *Location or detail drawings:* These show the positions occupied by the various spaces and elements of the building, together with dimensions. If no setting out drawing has been supplied, the site surveyor must prepare his own from these drawings and they must be checked carefully; e.g. to ensure that overall measurements agree with summed separate measurements, that calculated angles are correct, that heights agree, etc. Scales 1:50 or 1:100.

(d) *Specification:* A description of materials and workmanship, this may contain information relevant to setting out; e.g. depth of bed for drain lines.

(e) *Schedules:* Tabulated information on numerous and repetitive items, and again there may be relevant data; e.g. manhole positions, heights, invert levels.

(f) *Survey data:* Relates to the original site survey; e.g. location of traverse stations, permanent marks left on site in the original survey, bench mark data, etc.

8.2.3 INSPECTING THE SITE

This is as important as a survey recce, although for a different purpose. The nature of the site should be noted, and the presence of obstacles to ranging and measurement. The nature of the ground will affect the type of site markers to be used and the reliability of existing survey marks.

The area generally should be checked to verify agreement with the block and site plan details, and any changes notified.

Existing control markers and bench marks must be verified, and changes noted.

8.2.4 COMMON SETTING OUT TECHNIQUES

Setting out requires not only standard survey equipment and methods, but also a range of standard techniques not generally used in survey. These involve lining-in a point between two existing points (a more accurate version of ranging an intermediate point on a line in chain survey), extending an existing alignment, setting out an exact horizontal distance, setting out an exact horizontal angle (most often 90°), and setting out a particular (specified) height.

(a) Lining-in between two existing points.
If chain survey accuracy is adequate, as it may be on some preliminary or 'rough' location work, the methods described for this task in chaining may be used, as in §§ 2.2 and 2.6, but the accuracy of alignment may be only ± 50 mm.

For a higher standard of accuracy, as is generally required in setting out, a theodolite must be used for the lining-in, the method depending upon the intervisibility of the ends of the line joining the points and whether it is possible to set up the instrument over one of the points. Three cases arise, these being: End points intervisible, instrument can be set over one end; end points not intervisible, *or* instrument cannot be set over one end point; there is no intermediate point at which both end points are visible.

If the first case holds, carefully set up the theodolite over one end point, say A, then direct the telescope towards the other end point, say B, and bisect the target at B with the cross-hairs, finishing with all horizontal clamps applied. Despatch an assistant to the approximate position of the required intermediate point, and when he is roughly on line elevate or depress the telescope and re-focus. Looking through the telescope, direct the assistant to bring an arrow or similar fine mark exactly to the centre of the cross. If required, remove the temporary mark and place a suitable site marker, then check the alignment again with the theodolite. Repeat the process if other points are to be placed on the same alignment. Note that a piece of white card held behind the arrow assists contrast when visibility is poor.

If the ends are not intervisible, or the end points are not accessible, use chain survey techniques to locate the approximate position of an intermediate point (say C) from which both end points can be seen, and set the theodolite over C. On face left, sight on point A and bisect the target carefully, then transit the telescope and direct an assistant to place an arrow exactly on the collimation line in the vicinity of point B. (Except by accident, it is unlikely that the line defined by the telescope will coincide exactly with point B.) Keeping the instrument on face right, as it now is, turn to sight A again and bisect the target at A. Transit the telescope and direct the

assistant to place another mark on the telescope line in the vicinity of B. The two arrows will probably finish as shown in *Figure 8.5*, their mean position defining the alignment from A through the instrument station (any difference is due to collimation error).

Figure 8.5   To instrument at A

If the position indicated by the arrows is 100 mm (for example) from B (the 'near point' in *Figure 8.6*), and the theodolite is about halfway along the line, then by simple proportion the theodolite must be moved left by 50 mm to bring it onto the true alignment AB. Whether to move the tripod or simply to slide the instrument on the tripod head will depend upon the amount of

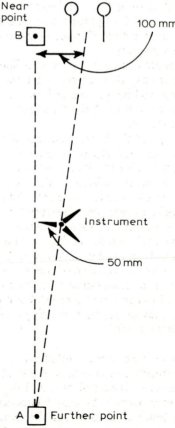

Figure 8.6

movement needed. When the instrument has been moved, repeat the process and continue repeating as needed. When the instrument is considered to be located on the alignment correctly, further points may be located as described in the first case above.

If the end points of a line are not intervisible, and there is no intermediate point from which both ends can be viewed (e.g. in a wooded area), traverse survey techniques as described in *Site Surveying 3* must be used.

(b) Extending an existing alignment
The typical problem here is where the alignment of existing site markers A and B in *Figure 8.7* is

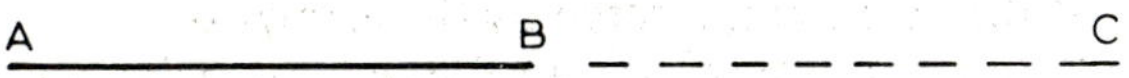

A _____ B _ _ _ _ _ _ _ C

Figure 8.7

to be extended to points such as C. Again, chain survey methods are available, but these are not generally suitable for setting out works.

The preferred method is to set up the theodolite over A, point towards B and carefully bisect the target at B, finishing with horizontal clamps applied, then re-focus the telescope and direct an assistant at C to place a mark exactly at the cross-hair intersection. Further points may be located in the same way.

If the theodolite cannot be set up over station A, or if it is not possible to sight the ground at C from A, say due to high ground at B or the ground falling from B to C, the theodolite must be set up over station B.

In this case, point the instrument carefully on the target at A, on face left, finishing with horizontal clamps applied, then transit the telescope and direct an assistant in the area of C to place a fine mark exactly at the cross-hair intersection. Still on face right (because the instrument was transitted), swing to point station A again, then transit and direct the positioning of another fine mark on the new alignment. A difference between the position of the two marks indicates instrument collimation error (assuming the instrument has been used correctly) and the mean position may be accepted as being on the extended alignment ABC.

(c) Setting out an exact horizontal distance
The typical problem here is to set out a point P, on a line joining two fixed points A and B, at a specified distance from A. Since it is impractical to define direction and distance simultaneously, this usually requires setting out the distance,

correcting the alignment, correcting the distance, and so on, repeating these in turn until the marker is exactly on line at the exact required distance.

The first step is to set up the theodolite over point A with the telescope directed on and bisecting the target at B, and points may be lined-in on the line AB as required (or on the line AB extended sometimes).

If the distance AP is less than one tape length, direct an assistant to pull out the required length of tape and hold the peg at the appropriate distance. After checking the peg alignment by sighting through the telescope, direct the assistant to drive the peg into the ground. Finally, measure the distance again carefully and make a fine mark on the peg, then adjust the alignment of the mark again.

If the distance AP is more than one tape length a more complex process is needed. Assuming that the distance AP is to be exactly 100.000 m, and that the theodolite has been set up as described above, measure out approximately 100 m with a synthetic tape and place a marker on the required alignment (checking alignment with the theodolite). If the temporary marker is designated as X, then it will be clear that the distance AX is unlikely to be exactly 100.000 m, but it will be fairly close to the required position. Measure the distance AX exactly, using the steel tape with the correct tension applied and correcting for slope, temperature and standardisation as necessary according to the accuracy specified.

If the corrected measured distance AX was found to be, say, 99.835 m, then this would indicate that the marker at X was some 0.165 m short of the required position P. The additional distance must be taped on (sometimes back, of course) from X, while checking the alignment with the theodolite, and the marker P located at the required distance along the line AB. As a final check, the length must be measured independently from P back towards A, applying all corrections as necessary. Alternatively, if the distance AB is known then the check may be carried out by measuring from P to B and comparing with the known distance.

If the additional distance XP (to be added or subtracted from AX) is large, then it may be necessary to apply corrections to this measurement also.

(d) Setting out an exact horizontal angle
For the great majority of buildings, the only

angles which require to be set out are angles of 90°. In a few cases other angles must be set out, but when using a theodolite the technique is the same whatever angular value is demanded. Here 90° will be assumed.

For lower accuracy work, a right angle may be set out with the steel tape, using the '3-4-5 triangle' method described in § 2.7. Again, on many small building jobs right angles can be set out quite satisfactorily using a large wooden right-angled triangle frame. For rough preliminary work, the optical square is useful, but it is not suitable for setting out buildings. For higher accuracy work the theodolite is the best instrument to use, although the site square instrument may be acceptable for smaller jobs.

To set out an angle by theodolite, the method is essentially an application of the simple reversal technique. Set the theodolite over the point on the line at which the right angle (or other desired angle) is to be raised, and level up as usual. On face left, sight onto a target placed on the far end of the line (i.e. the reference object) and read and book the horizontal circle. (Any specified zero may be set on the circle before sighting the RO, often a zero of 0°00′00″ is used.) Release the (upper) plate and telescope clamps and turn the instrument through 90° exactly (or other desired angle) and direct an assistant to place an arrow or other fine mark at the required distance on the alignment fixed by the telescope cross-hairs.

Release the (upper) plate and telescope clamps again, transit the telescope onto face right, turn to point the RO target again, read and book the horizontal circle reading. Release (upper) plate and telescope clamps, turn the instrument through 90° (or other desired angle) and direct an assistant to place another arrow or other fine mark at the required distance on the alignment fixed by the telescope cross-hairs. In theory the two arrow positions should coincide, but there will usually be a gap between them, its size depending upon the sighting distance and the instrument in use. If both sightings were considered to be reliable, take the mean position between the two arrows as being the required direction. The discrepancy, of course, is due to instrument error and small personal errors. If there is any doubt about the result, set a new zero on the circle and repeat the whole process as a check.

(e) Setting out a particular height.
It is normal practice to level from the nearest OSBM and establish TBMs on the site, these temporary bench marks being placed in positions which the surveyor may access easily yet will be well away from construction site traffic. TBMs may be steel rods set in concrete with the relevant information scribed in the concrete when it was wet, or any other suitable semi-permanent stable form of marker. The TBMs, are intended for the surveyor's use to avoid the need to level from the OSBM each time site levelling has to be done.

For the convenience of the construction team, however, additional height marker pegs are needed on site, in various positions and often with a specified height or level marked on the peg(s).

To set out such a peg with a specified height marked, first find the approximate level of the ground where the peg is to be placed, using ordinary levelling, then drive in a peg at the point, making sure that the top of the peg is higher than the specified level. Find the reduced level of the top of the peg, then mark a line on the side of the peg at the required height by scaling down from the top of the peg (*Figure 8.8*). As a check, place the staff zero against the

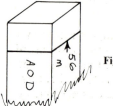

Figure 8.8

mark and ensure that the level is correct by reading the staff and calculating the reduced level of the mark.

8.2.5 SETTING OUT THE BUILDING PLAN DETAIL
Before commencing to set out the outline of the building plan, the site surveyor should have checked the appropriate drawings and other documents, looking for discrepancies and any 'lack of fit'. If any errors are found, they must be reported and corrections authorised before the detail setting out starts. In the interests of the survey principle of 'economy of accuracy', any specific requirements in respect of accuracies to be attained must be clarified also before starting work. As an example, it might be specified that the corners of a steel-framed building should agree with one another within a tolerance of 10 mm, yet the building as a whole need only be within 200 mm of its specified position — clearly in this case it would be wasteful to maintain the 10 mm tolerance in the building positioning as

well as in its relationship of its parts, since 'accuracy = money'.

For simple rectangular buildings generally, site markers locating the corners of the building are positioned by linear measurement with the steel tape in conjunction with rectangular offsets (angles raised as described in 'Setting out an exact horizontal angle', § 8.2.4(c)) and straights. This sounds like chain survey in reverse, but the accuracy required is much greater than that demanded in the preparation of plans. This 'inverse chain survey' method is employed because it is usually the method used by the design team in planning the site layout, though they work on a plan of the site and not on the site itself. Proposed new buildings are often planned to lie on the extension of existing alignments (a new building front in line with an existing building front, for example) or on lines parallel to, or on, an existing alignment (e.g. the face of a new building to lie on the line joining the corners of two existing buildings, or on a street 'building line').

In setting out site markers to fix corners of buildings etc., the following methods may be used to locate points on site:

(a) Linear measurements along the extension of an existing alignment. (Unless the positions of points can be checked by measurements from other features, measurements along alignment extensions should not generally exceed one tape length.)

(b) Linear measurements along a line between 'control' points (existing detail), the distance to a point fixed in this way being checked by measurement from both ends of the line.

(c) Minor control points may be fixed by locating the intersection of lines joining existing control points, then these new points may be used in (a) and (b) above.

(d) Minor control points (or actual site markers) may be fixed by taking three or more linear measurements from existing control points.

(e) Perpendicular offsets may be raised from points on the lines in (a) and (b), right angles being set out by theodolite or 3:4:5 triangle with the tape.

(f) Combinations of any of the methods detailed above.

The general approach is illustrated by the following two examples, but it must be appreciated that every setting out task is different and, except for the very simplest jobs, it is unlikely that any two independent surveyors would carry out the same job in exactly the same manner.

Example problem 1
Requirement:
To set out the simple rectangular building shown in *Figure 8.9*.

Given data:
(Abstracted from the contract documents) Size of the building, distance from adjacent building, front face to be on the line AB, as in *Figure 8.10*.

Suggested procedure:
To set out the works, and also to check the

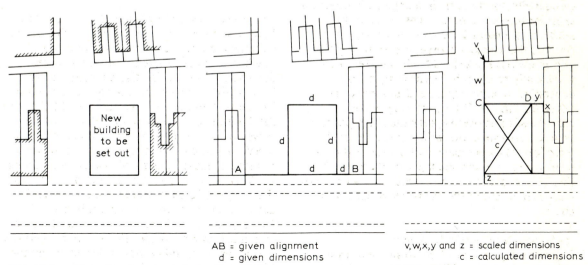

AB = given alignment
d = given dimensions

v, w, x, y and z = scaled dimensions
c = calculated dimensions

Figure 8.9   **Figure 8.10**   **Figure 8.11**

119

setting out with respect to itself and the surrounding detail.)

(a) Measure AB on site, scale the distance on the site plan; the two should agree within the tolerances stated in § 3.7.2. If no agreement, check site plan for possible survey errors, check drawing scale.

(b) After accepting or adjusting the site plan (as required) place two pegs on the line AB at the given distances to represent the front corners of the building. With the tape correctly positioned between A and B, insert nails in the pegs at the exact required distances from B.

(c) Set out the rear corner pegs by raising right angles at the front corner points (theodolite or other method) then measuring the given distances from the front pegs along the offsets. Again, pegs with a nail or other fine mark should be inserted.

(d) Calculate the theoretical lengths of the diagonals of the building, then measure the diagonals and CD on site — the actual and theoretical measurements should agree within $2\sqrt{L}$ mm, where L is the length concerned in metres. BS 5964 refers to 'Methods for setting out and measurement of buildings: permissible measuring deviations', but this generally deals with more complex sites (see *Site Surveying 3*).

(e) When satisfied that the building is correct within itself, check its location with respect to surounding detail. In this case, check that the site

measured and plan scaled values of v, w, x, y and z in *Figure 8.11* agree within the limits of tolerance mentioned in § 3.7.2.

Example problem 2
Requirement:
To set out the simple rectangular building shown in *Figure 8.12*.

Given data:
(Abstracted from the contract documents) Size of the building, and its general location as shown in *Figure 8.12*.

Suggested procedure:
(To set out the works, and also to check the setting out with respect to itself and the surrounding detail.)

(a) It is desirable in this case to intensify the 'control' since there are only five points of detail available. This will reduce the amount of taping required. As a first step, locate the intersected point I on the ground (*Figure 8.13*) using the optical square, and measure the distance IJ. The measured and scaled distances IJ should agree, thus checking the accuracy of the new control point and the five points of detail.

(b) Referring to *Figure 8.14*, carefully extend the building face AB on the site plan so that it meets the road at R and the line IJ at K. Using a set square (for an accurate right angle), extend the building side BC on the site plan to meet the hedge at point H.

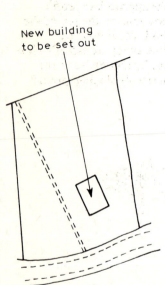

Figure 8.12

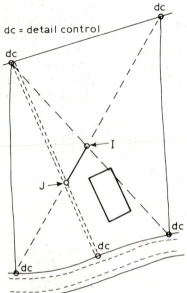

Figure 8.13

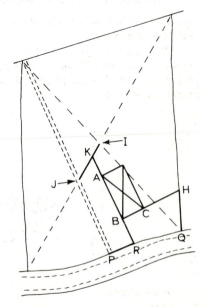

Figure 8.14

(c) Set out the points K and R on the ground, using scaled distances, then lay out the line KABR, obtaining the best fit. Scaled distances should agree with the tolerances in § 3.7.2 and the scaled and measured distances within $2\sqrt{L}$ mm again. Note that a slight movement of the point K along the line IJ and, to a lesser extent, of the point R along PQ, may extend or reduce the length of the line KABR.

(d) When satisfied with the line KABR insert pegs with nails at points A and B.

(e) Raise right angles at A and B (theodolite or other method), extending the right angle at B to give point H. Set out the building side lengths from A and B, locating pegs with nails for the rear corners of the building.

(f) Calculate the lengths of the diagonals, and measure the diagonals and the length of the rear face of the building between the nails. Calculated and actual measurements of these should again agree within $2\sqrt{L}$ mm, L being length in metres.

(g) When satisfied that the building is correct within itself, complete the check with the surrounding detail by comparing the scaled and measured distances CH and HQ.

8.3 Field procedure — setting out profile boards

Aim: *The student should be able to describe and undertake the positioning of profiles and the level datum for a building.*

Figure 8.15 illustrates a partially completed task, and what is required.

A, B, C and D represent pegs, placed at the proposed locations of the corners of a building, positioned and checked as described in the preceding section.

P, Q, R and S represent corner profiles, and on these will be marked the extended alignments (the 'straights') of the faces of the walls and foundations, the marks being used by the construction team to control the actual positioning of these elements as the work proceeds. The profiles must remain in position until all the construction on the alignments they define has been completed, therefore they must be placed clear of all likely construction and traffic, and they may have to straddle excavations such as for drainage works.

The profiles are set out horizontally, often at a specified height between 300 and 450 mm above existing ground level, and related to a known datum height such as ground floor level. The pegs and boards used must be robust, typical measurements being given in *Figure 8.3*

The procedure to set up the profiles in *Figure 8.15* may be summarised as follows:

(a) At each corner, drive four pegs (or stakes, depending on the ground) to act as supports for the horizontal boards.

(b) Mark the agreed datum height on the inside edge of each peg.

(c) Nail the horizontal boards to the pegs or stakes, with the tops of the boards level with the datum height marks, and check with the canoe level that they are horizontal.

(d) If required, mark the datum height on each board (*Figure 8.16*).

(e) Extend the alignment of the pegs A and B in each direction to cut the appropriate profile boards, marking the positions on the top of the boards with pencil. This is often done by stretching a builder's line along the extended line AB, but it is better done by theodolite or site square, particularly on steeply sloping ground.

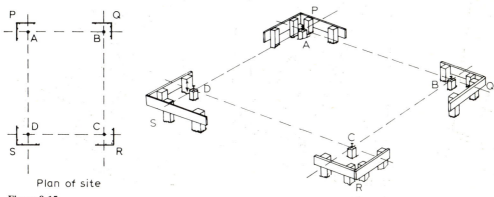

Plan of site

Figure 8.15

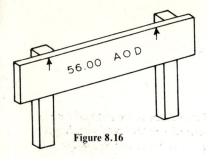

Figure 8.16

(f) Extend the alignments BC, CD and DA to the respective profile boards in the same way as for AB.

(g) Measure from the pencil marks on the board tops to locate the alignments of the inner and outer faces of walls, foundations, etc., making suitable fine pencil marks. (Avoid making too many marks or confusion may result.) When satisfied that all marks are in the correct positions, replace them by fine saw cuts in the board top, or small nails.

(h) If the building is not a simple rectangle, as in *Figure 8.17,* place any necessary additional profile boards.

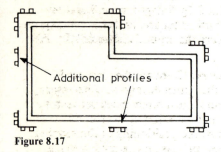

Additional profiles

Figure 8.17

8.4 Controlling line and height
Aim: *The student should be able to explain how profiles are used with a traveller to control excavation and foundation levels.*

8.4.1 CONTROLLING LINE
When the corner pegs A, B, C and D in *Figure 8.15* have been removed after fixing the profiles, ground positions may be located by plumbing down (plumb-bob and cord) from the builder's lines strung between marks on the profiles (*Figure 8.18*). Wall and foundation corners may be located by plumbing down from the appropriate intersections of string lines in the same way. In some cases, however, it will be found best to set up the theodolite over a profile board mark and direct the telescope along the

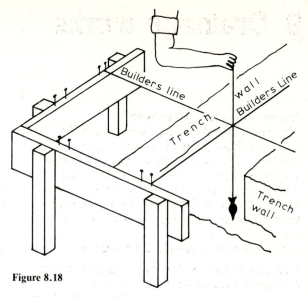

Figure 8.18

alignment required to the next profile, then line markers in as needed.

8.4.2 CONTROLLING HEIGHT
Excavation depth also is controlled from the stretched builder's lines, since these are attached to the top of the profiles and the profiles have been fixed at a specified height. It would be possible to measure down from the builder's line with a tape, but it is difficult, and instead the preferred method is to make a *traveller* of the required length (*Figure 8.19*) and this can be used

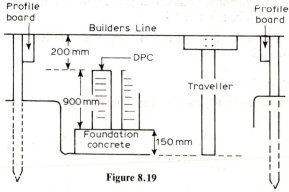

Figure 8.19

by one man. While carrying out trench excavation the site operative may place the traveller on the bottom of the trench, alongside the stretched line, and when the top of the traveller is level with the line the trench depth is correct.

As an independent check, and to control the finished foundation level, the site surveyor may then insert pegs in, and along the edges of, the completed excavation, using a level and staff.

9 Drainage works

Introduction and procedure

The setting out of drainage work is usually relatively simple, being concerned again with the control of line and height. Despite the apparent simplicity, however, very careful checks must be carried out, since it can be very expensive to replace, for example, a deep sewer which has been laid on the wrong line or at the wrong depth.

The operations typically involved are:

(a) Examine and check relevant documents for plan position and height and gradient information, ensuring that gradients and heights shown are compatible.

(b) Inspect the site, check the position of relevant existing manholes and the heights, and prove the site drawings. Note that information on existing old manholes and drains is often inaccurate, and problems may arise in the need to cross existing services lines.

(c) Locate the proposed manhole positions by the usual setting out methods, scaling from the drawings and measuring out on the ground, and place appropriate marker pegs and profiles for plan and height control.

(d) When complete, check all pegs and profiles carefully.

Example problem

As shown in *Figure 9.1* , the positions of manholes (MH) 2 and 3 are to be located on the ground, given a plan of the area and the design details of distance between manholes, heights, gradients, etc.

Suggested procedure:

(a) *On the site plan:* Extend the alignment of the two manholes so as to cut the existing fence lines at points A and B. Draw the line XY, cutting the line of the manholes at Z and the field boundary at W. Scale off the distances from A through MH2, Z and MH3 to B. Scale the distances from A and B to the nearest corners of the field at C and D respectively, and scale the distances ZW and CW.

(b) *In the field:* Set out the scaled distances CA, CW and DB from C and D respectively, inserting ranging rods at A, W and B. Locate the intersected point Z by lining-in on AB and on XW using the optical square, and insert a ranging rod at Z. Measure the ground distances ZW, AB and AZ.

(c) *Check the work:* Compare the measured distances with the scaled distances. If necessary, provided that no large discrepancies between the plan and the ground are found, correct or judiciously adjust the positioning of the ranging rods for the 'best fit'. As an example, moving either or both of the rods at A and B towards the road will have the effect of shortening the distances AB and ZW, and vice versa; moving

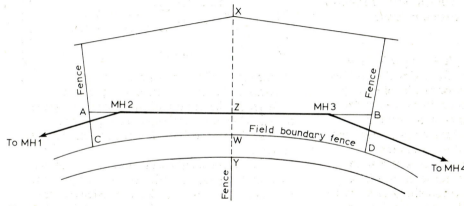

Figure 9.1

only rod A towards the road will shorten AZ also, but ZB will hardly be affected, and so on.

(d) When satisfied, locate the manholes on the ground by setting out the scaled distances A-MH2 and B-MH3, placing pegs, and measuring MH2-MH3 as a check. profiles may be set as needed, and heights transferred from BMs as necessary.

9.1 The sight rail and traveller
Aim: *The student should be able to identify and describe the invert of a drain, a sight rail and a traveller.*

The depth and alignment of drains and trenches can be controlled by the use of sight rails and travellers, both of which are described in § 8.1.

A sight rail is merely a special type of profile board erected over a proposed trench line at suitable intervals. In setting out a rectangular building, a profile is used to control alignment and may also be used to control depth by using a traveller in conjunction with a builder's line stretched between profile marks. The sight rail is used similarly but without the stretched builder's line, the operator merely sights from one sight rail to the next, and when the traveller is held at an intermediate point in the drain line the trench is at the correct depth when the top of the traveller just touches the sight line from one rail to the next.

Since sight rails are primarily used for checking excavation depth, they should be constructed at a convenient viewing height (about 1.5 m above the ground) for ease of use, unlike building profile boards which are generally placed quite close to ground level. As with building profiles, alignment marks for the trench and pipe may be made on the site rails and alignment can be controlled by sighting from mark to mark along the line, an

assistant plumbing down as required.

Typical sight rails in use are shown in *Figure 9.2*. Forms of sight rail may vary according to whether mechanical or hand excavation is being used, and it may be necessary to vary the length of the traveller. It will be evident that sight rails with two supports are likely to be more stable.

Figure 9.3 shows a traveller in use. The final height to be controlled by the use of the traveller is the *invert level,* the lowest point on the inside surface of the drain pipe (*Figure 9.4*). The drainage schedule may give tabulated information on the manholes, including the cover height and the invert levels, and the appropriate length of traveller should be calculated as described in § 9.2.

9.2 Calculations for sight rails
Aim: *The student should be able to calculate a suitable length for a traveller and the reduced levels of sight rails, given the relevant drawings.*

For a straight drain line, the minimum number of sight rails required for controlling the work is two, but as a general rule three should be provided. The three rails should be established separately, the third acting as a check on the others, and the existence of a third rail provides some security if one rail is accidentally (or otherwise) destroyed or damaged. The most suitable arrangement is usually to provide one rail at the centre of the drain length and the others on the extended alignment, beyond the manhole at each end. It is preferable to keep the sight rails clear of manhole positions in order to avoid obstructing construction and also to minimise damage to the sight rails.

Figure 9.5 illustrates an example problem, with the information which should be obtained from

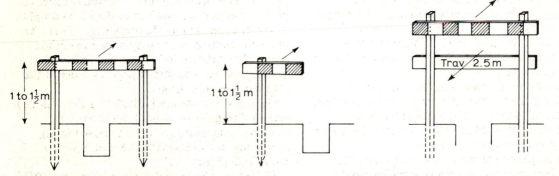

Figure 9.2 The surveyor stands at the painted side of the profile to view the traveller, as indicated by the arrowheads on the sketches. Note: the length of the traveller may be written on the reverse side

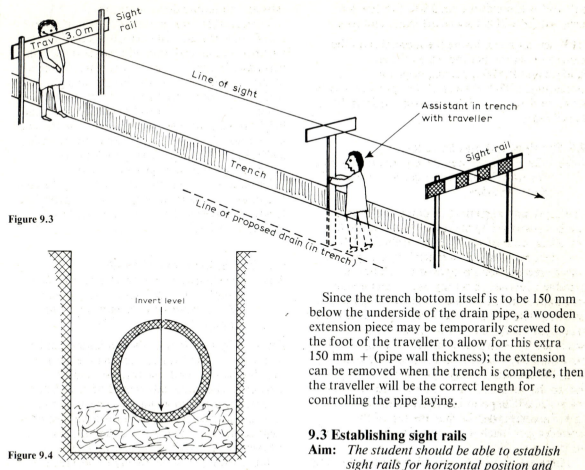

Figure 9.3

Figure 9.4

Invert level

the contract documents plus the reduced levels as calculated from levelling observations carried out by the site surveyor.

Notes on the example:

Sight rail to be between 1 and 1.5 m above existing ground level. The end sight rails are to be placed 3 m beyond the manholes, and with a gradient of 1 in 60 this means that the theoretical invert levels at the end sight rails will be 0.05 m higher or lower than the manhole invert levels. The centre invert level will be the mean of the end manhole invert levels.

The maximum and minimum length of traveller is the difference in height of the previous two sets of figures. An acceptable traveller height would be 4.25 m which is within the above maximum and minimum range.

The height of the top of the sight rail is the invert level plus the acceptable traveller height. The final figure is the height of the top of the sight rail above the ground level peg.

Since the trench bottom itself is to be 150 mm below the underside of the drain pipe, a wooden extension piece may be temporarily screwed to the foot of the traveller to allow for this extra 150 mm + (pipe wall thickness); the extension can be removed when the trench is complete, then the traveller will be the correct length for controlling the pipe laying.

9.3 Establishing sight rails

Aim: *The student should be able to establish sight rails for horizontal position and depth control for a straight drain between manholes.*

The manhole locating pegs first set out will disappear when construction work starts, thus, as shown in the preceding section, it is normal practice to establish pegs for sight rail positions some 3 m beyond the manholes on the extended drain alignment, so that the sight rails will be well clear of the works. It is also common practice to place offset pegs at, say 3 m to the left and right of the original manhole locating peg, so that the manhole centre position may be re-established as required and the drain line can be re-aligned. It is sensible also to place offset pegs for a peg at the centre of the drain length.

Sight rails may be established over the sight rail pegs by fixing the horizontal rail to vertical supports, controlling the height above the peg with a tape and checking horizontality with a canoe level (*Figure 9.6*).

If the trench is to be dug by machine, an offset profile must be set up on the side of the trench opposite to that on which the soil is to be

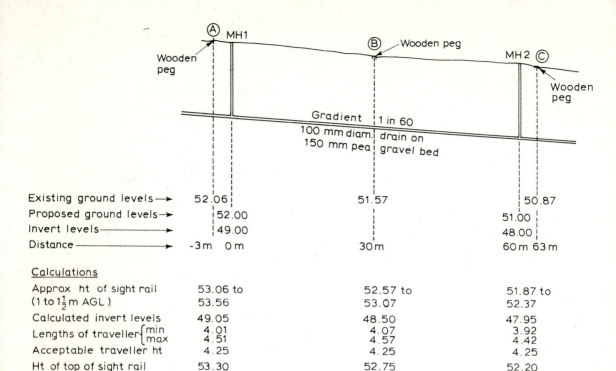

Existing ground levels →	52.06	51.57	50.87
Proposed ground levels →	52.00		51.00
Invert levels →	49.00		48.00
Distance →	-3m 0m	30m	60 m 63 m

Calculations

Approx ht of sight rail (1 to 1½ m AGL)	53.06 to 53.56	52.57 to 53.07	51.87 to 52.37
Calculated invert levels	49.05	48.50	47.95
Lengths of traveller {min max	4.01 4.51	4.07 4.57	3.92 4.42
Acceptable traveller ht	4.25	4.25	4.25
Ht of top of sight rail and above peg	53.30 1.24	52.75 1.18	52.20 1.33

Figure 9.5

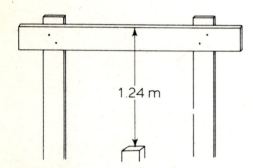

1.24 m

Figure 9.6 The height of 1.24 m is obtained from calculations at peg A in Figure 9.5

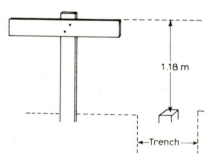

1.18 m

←Trench→

Figure 9.7 The height of 1.18 m is obtained from calculations at peg B in Figure 9.5

dumped (*Figure 9.7*). This situation may perhaps be best tackled by using a level and staff again.

An offset profile will require the use of an inverted L-shaped traveller (*Figure 9.8*).

When the third sight rail has been established, the top horizontal bar of all three profiles should lie on the same gradient, that is they should define a single straight sight line. If they do not do so, then there is an error(s) in the heighting of the pegs, in the calculations, or in the actual placing of the sight rails — all should be checked as necessary until the error has been located and eliminated.

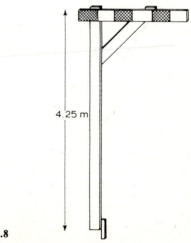

4.25 m

Figure 9.8